Statistics

Data: Kids, Cats, and Ads

Grade 5

Also appropriate for Grade 6

Andee Rubin
Jan Mokros
Rebecca Corwin
Susan Friel

Developed at TERC, Cambridge, Massachusetts

Dale Seymour Publications®

The *Investigations* curriculum was developed at TERC (formerly Technical Education Research Centers) in collaboration with Kent State University and the State University of New York at Buffalo. The work was supported in part by National Science Foundation Grant No. MDR-9050210. TERC is a nonprofit company working to improve mathematics and science education. TERC is located at 2067 Massachusetts Avenue, Cambridge, MA 02140.

This project was supported, in part, by the
National Science Foundation
Opinions expressed are those of the authors and not necessarily those of the Foundation

This book is published by Dale Seymour Publications®, an imprint of the Alternative Publishing Group of Addison-Wesley Publishing Company.

Managing Editor: Catherine Anderson
Series Editor: Beverly Cory
Manuscript Editor: Nancy Tune
Consulting Editor: Priscilla Cox Samii
ESL Consultant: Nancy Sokol Green
Production/Manufacturing Director: Janet Yearian
Production/Manufacturing Supervisor: Karen Edmonds
Production/Manufacturing Coordinator: Barbara Atmore
Design Manager: Jeff Kelly
Design: Don Taka
Illustrations: DJ Simison, Carl Yoshihara
Cover: Bay Graphics
Composition: Publishing Support Services

Apple and Macintosh are registered trademarks of Apple Computer, Inc.
Microsoft is a registered trademark and Windows is a trademark of Microsoft Corporation.
Tabletop Sr. is a trademark of TERC.

This book is based on *Statistics: Prediction and Sampling* by Rebecca B. Corwin and Susan N. Friel, which was originally published by Dale Seymour Publications® as part of the series *Used Numbers: Real Data in the Classroom*, © 1990 by Dale Seymour Publications®.

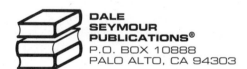

DALE SEYMOUR PUBLICATIONS®
P.O. BOX 10888
PALO ALTO, CA 94303

Order number DS21430

ISBN 0-86651-994-7

1 2 3 4 5 6 7 8 9 10-ML-99 98 97 96 95

T E R C

Principal Investigator   Susan Jo Russell

Co-Principal Investigator   Cornelia C. Tierney

Director of Research and Evaluation   Jan Mokros

Curriculum Development

Joan Akers
Michael T. Battista
Mary Berle-Carman
Douglas H. Clements
Karen Economopoulos
Claryce Evans
Marlene Kliman
Cliff Konold
Jan Mokros
Megan Murray
Ricardo Nemirovsky
Tracy Noble
Andee Rubin
Susan Jo Russell
Margie Singer
Cornelia C. Tierney

Evaluation and Assessment

Mary Berle-Carman
Jan Mokros
Andee Rubin
Tracey Wright

Teacher Support

Kabba Colley
Karen Economopoulos
Anne Goodrow
Nancy Ishihara
Liana Laughlin
Jerrie Moffett
Megan Murray
Margie Singer
Dewi Win
Virginia Woolley
Tracey Wright
Lisa Yaffee

Administration and Production

Irene Baker
Amy Catlin
Amy Taber

Cooperating Classrooms for This Unit

Sarah Napier
Fayerweather Street School, Cambridge, MA
Barbara Fox
Cambridge Public Schools, Cambridge, MA
Dorothy Spahr
Wellesley Public Schools, Wellesley, MA
Reina Huerta
John Wolfe
New York City Public Schools, New York, NY
Christine Fuentes
Diane Pearson
Clarke County Public Schools, GA

Technology Development

Douglas H. Clements
Julie Sarama

Video Production

David A. Smith
Judy Storeygard

Consultants and Advisors

Deborah Lowenberg Ball
Marilyn Burns
Mary Johnson
James J. Kaput
Mary M. Lindquist
Leslie P. Steffe
Grayson Wheatley

Graduate Assistants

Kent State University:
Richard Aistrope, Kathryn Battista,
Caroline Borrow, William Hunt
State University of New York at Buffalo:
Jeffery Barrett, Julie Sarama,
Sudha Swaminathan, Elaine Vukelic
Harvard Graduate School of Education:
Dan Gillette, Irene Hall

CONTENTS

Teacher Notes

Investigations in Number, Data, and Space is a K–5 mathematics curriculum with four major goals:

- to offer students meaningful mathematical problems
- to emphasize depth in mathematical thinking rather than superficial exposure to a series of fragmented topics
- to communicate mathematics content and pedagogy to teachers
- to substantially expand the pool of mathematically literate students

The *Investigations* curriculum embodies an approach radically different from the traditional textbook-based curriculum. At each grade level, it consists of a set of separate units, each offering 2–6 weeks of work. These units of study are presented through investigations that involve students in the exploration of major mathematical ideas.

Approaching the mathematics content through investigations helps students develop flexibility and confidence in approaching problems, fluency in using mathematical skills and tools to solve problems, and proficiency in evaluating their solutions. Students also build a repertoire of ways to communicate about their mathematical thinking, while their enjoyment and appreciation of mathematics grows.

The investigations are carefully designed to invite all students into mathematics—girls and boys, diverse cultural, ethnic, and language groups, and students with different strengths and interests. Problem contexts often call on students to share experiences from their family, culture, or community. The curriculum eliminates barriers—such as work in isolation from peers, or emphasis on speed and memorization—that exclude some students from participating successfully in mathematics. The following aspects of the curriculum ensure that all students are included in significant mathematics learning:

- Students spend time exploring problems in depth.
- They find more than one solution to many of the problems they work on.

- They invent their own strategies and approaches, rather than relying on memorized procedures.
- They choose from a variety of concrete materials and appropriate technology, including calculators, as a natural part of their everyday mathematical work.
- They express their mathematical thinking through drawing, writing, and talking.
- They work in a variety of groupings—as a whole class, individually, in pairs, and in small groups.
- They move around the classroom as they explore the mathematics in their environment and talk with their peers.

While reading and other language activities are typically given a great deal of time and emphasis in elementary classrooms, mathematics often does not get the time it needs. If students are to experience mathematics in depth, they must have enough time to become engaged in real mathematical problems. We believe that a minimum of five hours of mathematics classroom time a week—about an hour a day—is critical at the elementary level. The plan and pacing of the *Investigations* curriculum is based on that belief.

For further information about the pedagogy and principles that underlie these investigations, see the Teacher Notes throughout the units and the following books:

- *Implementing the* Investigations in Number, Data, and Space™ *Curriculum*
- *Beyond Arithmetic: Changing Mathematics in the Elementary Classroom*

The *Investigations* curriculum is presented through a series of teacher books, one for each unit of study. These books not only provide a complete mathematics curriculum for your students, they offer materials to support your own professional development. You, the teacher, are the person who will make this curriculum come alive in the classroom; the book for each unit is your main support system.

While reproducible resources for students are provided, the curriculum does not include student books. Students work actively with objects and experiences in their own environment and with a variety of manipulative materials and technology, rather than with workbooks and worksheets filled with problems. We also make extensive use of the overhead projector as a way to present problems, to focus group discussion, and to help students share ideas and strategies. If an overhead projector is available, we urge you to try it as suggested in the investigations.

Ultimately, every teacher will use these investigations in ways that make sense for his or her particular style, the particular group of students, and the constraints and supports of a particular school environment. We have tried to provide with each unit the best information and guidance for a wide variety of situations, drawn from our collaborations with many teachers and students over many years. Our goal in this book is to help you, as a professional educator, implement this mathematics curriculum in a way that will give all your students access to mathematical power.

Investigation Format

The opening two pages of each investigation help you get ready for the student work that follows. Here you will read:

What Happens—a synopsis of each session or block of sessions.

Mathematical Emphasis—the most important ideas and processes students will encounter in this investigation.

What to Plan Ahead of Time—materials to gather, student sheets to duplicate, transparencies to make, and anything else you need to do before starting.

Examining Cats

What Happens

Session 1: A Set of Cats Students receive a set of Cat Cards, each of which provides data on one cat: weight, length, age, and gender as well as eye, pad, and fur color. In small groups, students focus on one particular variable as they make representations describing this group of cats.

Session 2: Associations: Are Female Cats Shorter? Students consider associations between two variables, such as length and gender, and determine how two variables might "go together."

Session 3 (Excursion): Using the Computer to Investigate Cats Students add their own cat data to a computer database and use it to explore questions about the relationships between variables.

Mathematical Emphasis

- Collecting and examining data that involve more than one variable
- Making representations of numerical and categorical variables
- Using fractions and percentages to understand categorical and numerical data
- Exploring the way two variables in a data set might be related
- Learning to enter and analyze data in a computer database

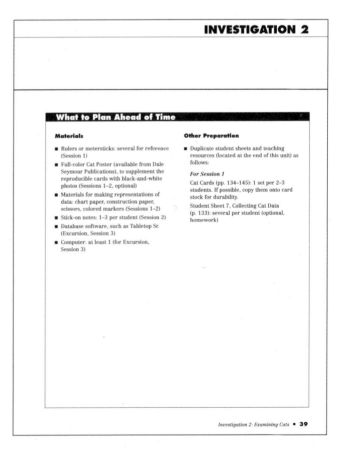

What to Plan Ahead of Time

Materials

- Rulers or metersticks: several for reference (Session 1)
- Full-color Cat Poster (available from Dale Seymour Publications), to supplement the reproducible cards with black-and-white photos (Sessions 1–2, optional)
- Materials for making representations of data: chart paper, construction paper, scissors, colored markers (Sessions 1–2)
- Stick-on notes: 1–3 per student (Session 2)
- Database software, such as Tabletop Sr. (Excursion, Session 3)
- Computer: at least 1 (for Excursion, Session 3)

Other Preparation

- Duplicate student sheets and teaching resources (located at the end of this unit) as follows:

  For Session 1

  Cat Cards (pp. 134–145): 1 set per 2–3 students. If possible, copy them onto card stock for durability.

  Student Sheet 7, Collecting Cat Data (p. 133): several per student (optional, homework)

Sessions   Within an investigation, the activities are organized by class session, a session being a one-hour math class. Sessions are numbered consecutively through an investigation. Often several sessions are grouped together, presenting a block of activities with a single major focus.

When you find a block of sessions presented together—for example, Sessions 1, 2, and 3—read through the entire block first to understand the overall flow and sequence of the activities. Make some preliminary decisions about how you will divide the activities into three sessions for your class, based on what you know about your students. You may need to modify your initial plans as you progress through the activities, and you may want to make notes in the margins of the pages as reminders for the next time you use the unit.

Be sure to read the Session Follow-Up section at the end of the session block to see what homework assignments and extensions are suggested as you make your initial plans.

While you may be used to a curriculum that tells you exactly what each class session should cover, we have found that the teacher is in a better position to make these decisions. Each unit is flexible and may be handled somewhat differently by every teacher. While we provide guidance for how many sessions a particular group of activities is likely to need, we want you to be active in determining an appropriate pace and the best transition points for your class.

Ten-Minute Math   At the beginning of some sessions, you will find Ten-Minute Math activities. These are designed to be used in tandem with the investigations, but not during the math hour. Rather, we hope you will do them whenever you have a spare 10 minutes—maybe before lunch or recess, or at the end of the day.

Ten-Minute Math offers practice in key concepts, but not always those being covered in the unit. For example, in a unit on using data, Ten-Minute Math might revisit geometric activities done earlier in the year. Complete directions for the suggested activities are included at the end of each unit. A compilation of Ten-Minute Math activities is also available as a separate book.

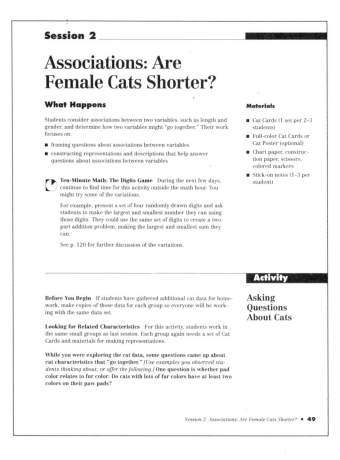

Session 2

Associations: Are Female Cats Shorter?

What Happens

Students consider associations between two variables, such as length and gender, and determine how two variables might "go together." Their work focuses on:

- framing questions about associations between variables
- constructing representations and descriptions that help answer questions about associations between variables

Ten-Minute Math: The Digits Game   During the next few days, continue to find time for this activity outside the math hour. You might try some of the variations.

For example, present a set of four randomly drawn digits and ask students to make the largest and smallest number they can using those digits. They could use the same set of digits to create a two-part addition problem, making the largest and smallest sum they can.

See p. 120 for further discussion of the variations.

Materials

- Cat Cards (1 set per 2–3 students)
- Full-color Cat Cards or Cat Poster (optional)
- Chart paper, construction paper, scissors, colored markers
- Stick-on notes (1–3 per student)

Activity

Before You Begin   If students have gathered additional cat data for homework, make copies of those data for each group so everyone will be working with the same data set.

Looking for Related Characteristics   For this activity, students work in the same small groups as last session. Each group again needs a set of Cat Cards and materials for making representations.

While you were exploring the cat data, some questions came up about cat characteristics that "go together." *[Use examples you observed students thinking about, or offer the following.]* One question is whether pad color relates to fur color: Do cats with lots of fur colors have at least two colors on their paw pads?

Asking Questions About Cats

Session 2: Associations: Are Female Cats Shorter? ▪ **49**

Activities   The activities include pair and small-group work, individual tasks, and whole-class discussions. In any case, students are seated together, talking and sharing ideas during all work times. Students most often work cooperatively, although each student may record work individually.

Choice Time   In some units, some sessions are structured with activity choices. In these cases, students may work simultaneously on different activities focused on the same mathematical ideas. Students choose which activities they want to do, and they cycle through them.

You will need to decide how to set up and introduce these activities and how to let students make their choices. Some teachers present them as station activities, in different parts of the room. Some list the choices on the board as reminders or have students keep their own lists.

Excursions   Some of the investigations in this unit include *excursions*—activities that could be omitted without harming the integrity of the unit. This is one way of dealing with the overabundance of fascinating mathematics to be studied—much more than a class has time to explore in any one year. Excursions give you the flexibility to make different choices from year to year. For example, you might do the excursion in *Data: Kids, Cats, and Ads* this year, but another year, try the excursions in another unit.

Tips for the Linguistically Diverse Classroom

At strategic points in each unit, you will find concrete suggestions for simple modifications of the teaching strategies to encourage the participation of all students. Many of these tips offer alternative ways to elicit critical thinking from students at varying levels of English proficiency, as well as from other students who find it difficult to verbalize their thinking.

The tips are supported by suggestions for specific vocabulary work to help ensure that all students can participate fully in the investigations. The Preview for the Linguistically Diverse Classroom (p. 13) lists important words that are assumed as part of the working vocabulary of the unit. Second-language learners will need to become familiar with these words in order to understand the problems and activities they will be doing. These terms can be incorporated into students' second-language work before or during the unit. Activities that can be used to present the words are found in the appendix, Vocabulary Support for Second-Language Learners (p. 124).

In addition, ideas for making connections to students' language and cultures, included on the Preview page, help the class explore the unit's concepts from a multicultural perspective.

Session Follow-Up

Homework   Homework is suggested on a regular basis in the grade 5 units. The homework may be used for (1) review and practice of work done in class; (2) preparation for activities coming up—for example, collecting data for a class project; or (3) involving and informing family members.

Some units in the *Investigations* curriculum have more homework than others, simply because it makes sense for the mathematics that's going on. Other units rely on manipulatives that most students won't have at home, making homework diffi-

cult. In any case, homework should always be directly connected to the investigations in the unit, or to work in previous units—never sheets of problems just to keep students busy.

Extensions   These follow-up activities are opportunities for some or all students to explore a topic in greater depth or in a different context. They are not designed for "fast" students; mathematics is a multifaceted discipline, and different students will want to go further in different investigations. Look for and encourage the sparks of interest and enthusiasm you see in your students, and use the extensions to help them pursue these interests.

Family Letter   A letter that you can send home to students' families is included with the blackline masters for each unit. We want families to be informed about the mathematics work in your classroom; they should be encouraged to participate in and support their children's work. A reminder to send home the letter appears in one of the early investigations. These letters are also available separately in Spanish, Vietnamese, Cantonese, Hmong, and Cambodian.

Materials

A complete list of the materials needed for the unit is found on p. 11. Some of these materials are available in a kit for the *Investigations* grade 5 curriculum. Individual items can also be purchased as needed from school supply stores and dealers.

In an active mathematics classroom, certain basic materials should be available at all times: interlocking cubes, pencils, unlined paper, graph paper, calculators, things to count with, and measuring tools. Some activities in this curriculum require scissors and glue sticks or tape. Stick-on notes and large paper are also useful materials throughout.

So that students can independently get what they need at any time, they should know where these materials are kept, how they are stored, and how they are to be returned to the storage area. For example, interlocking cubes are best stored in towers of ten; then, whatever the activity, they should be returned to storage in groups of ten at the end of the hour. You'll find that establishing such routines at the beginning of the year is well worth the time and effort.

Student Sheets and Teaching Resources

Reproducible pages to help you teach the unit are found at the end of this book. These include masters for making overhead transparencies and other teaching tools, as well as student recording sheets.

Many of the field-test teachers requested more sheets to help students record their work, and we have tried to be responsive to this need. At the same time, we think it's important that students find their own ways of organizing and recording their work. They need to learn how to explain their thinking with both drawings and written words, and how to organize their results so someone else can understand them.

To ensure that students get a chance to learn how to represent and organize their own work, we deliberately do not provide student sheets for every activity. We recommend that your students keep a mathematics notebook or folder so that their work, whether on reproducible sheets or their own paper, is always available to them for reference.

Help for You, the Teacher

Because we believe strongly that a new curriculum must help teachers think in new ways about mathematics and about their students' mathematical thinking processes, we have included a great deal of material to help you learn more about both.

About the Mathematics in This Unit This introductory section (p. 12) summarizes for you the critical information about the mathematics you will be teaching. This will be particularly valuable to teachers who are accustomed to a traditional textbook-based curriculum.

Teacher Notes These reference notes provide practical information about the mathematics you are teaching and about our experience with how students learn. Many of the notes were written in response to actual questions from teachers, or to discuss important things we saw happening in the field-test classrooms. Some teachers like to read them all before starting the unit, then review them as they come up in particular investigations.

Dialogue Boxes Sample dialogues demonstrate how students typically express their mathematical ideas, what issues and confusions arise in their thinking, and how some teachers have guided class discussions. These dialogues are based on the extensive classroom testing of this curriculum; many are word-for-word transcriptions of recorded class discussions. They are not always easy reading; sometimes it may take some effort to unravel what the students are trying to say. But this is the value of these dialogues; they offer good clues to how your students may develop and express their approaches and strategies, helping you prepare for your own class discussions.

Where to Start You may not have time to read everything the first time you use this unit. As a first-time user, you will likely focus on understanding the activities and working them out with your students. Read completely through each investigation before starting to present it.

When you next teach this same unit, you can begin to read more of the background. Each time you present this unit, you will learn more about how your students understand the mathematical ideas. The first-time user of *Data: Kids, Cats, and Ads* should read the following:

- About the Mathematics in This Unit (p. 12)
- Teacher Note: Finding Medians and Other Fractional Parts of Data Sets (p. 26)
- Teacher Note: Ways of Exploring and Representing Associations (p. 52)
- Teacher Note: Whys and Hows of Sampling (p. 74)

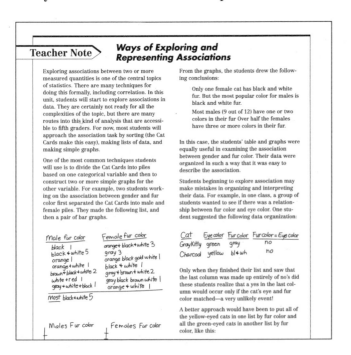

Teacher Checkpoints As a teacher of the *Investigations* curriculum, you observe students daily, listen to their discussions, look carefully at their work, and use this information to guide your teaching. We have designated Teacher Checkpoints as natural times to get an overall sense of how your class is doing in the unit.

The Teacher Checkpoints provide a time for you to pause and reflect on your teaching plan while observing students at work in an activity. These sections offer tips on what you should be looking for and how you might adjust your pacing. Are most students fluent with strategies for solving a particular kind of problem? Are they just starting to formulate good strategies? Or are they still struggling with how to start?

Depending on what you see as the students work, you may want to spend more time on similar problems, change some of the problems to use smaller numbers, move quickly to more challenging material, modify subsequent activities for some students, work on particular ideas with a small group, or pair students who have good strategies with those who are having more difficulty.

In *Data: Kids, Cats, and Ads* you will find three Teacher Checkpoints:

> Investigation 1, Session 4:
> The Mystery Balancers (p. 35)
>
> Investigation 2, Session 1:
> Representing Cat Data (p. 42)
>
> Investigation 4, Session 1:
> How Much of the Page Is Ads? (p. 84)

Embedded Assessment Activities Use the built-in assessments included in this unit to help you examine the work of individual students, figure out what it means, and provide feedback. From the students' point of view, the activities you will be using for assessment are no different from any others; they don't look or feel like traditional tests.

These activities sometimes involve writing and reflecting, at other times a brief interaction between student and teacher, and in still other instances the creation and explanation of a product.

In *Data: Kids, Cats, and Ads* you will find one assessment activity:

> Investigation 5, Sessions 3–5:
> What We Recommend (p. 113)

Teachers find the hardest part of the assessment to be interpreting their students' work. If you have used a process approach to teaching writing, you will find our mathematics approach familiar. To help with interpretation, we provide guidelines and questions to ask about the students' work. In some cases we include a Teacher Note with specific examples of student work and a commentary on what it indicates. This framework can help you determine how your students are progressing.

As you evaluate students' work, it's important to remember that you're looking for much more than the "right answer." You'll want to know what their strategies are for solving the problem, how well these strategies work, whether they can keep track of and logically organize an approach to the problem, and how they make use of representations and tools to solve the problem.

Ongoing Assessment Good assessment of student work involves a combination of approaches. Some of the things you might do on an ongoing basis include the following:

- **Observation** Circulate around the room to observe students as they work. Watch for the development of their mathematical strategies, and listen to their discussions of mathematical ideas.

- **Portfolios** Ask students to document their work, in journals, notebooks, or portfolios. Periodically review this work to see how their mathematical thinking and writing are changing. Some teachers have students keep a notebook or folder for each unit, while others prefer one mathematics notebook or a portfolio of selected work for the entire year. Take time at the end of each unit for students to choose work for their portfolios. You might also have them write about what they've learned in the unit.

Data: Kids, Cats, and Ads

OVERVIEW

Content of This Unit This unit builds on prior experience with collecting, graphing, and interpreting data, providing students with additional tools for data analysis. Students have several opportunities to compare two data sets, using "typical" values such as the median and also calculating what fraction of each data set is above or below a particular value. Sampling is introduced, and students begin to see how samples can be informative without providing all the information a population would. The unit culminates with a student-designed survey of a sample of their school to explore data about playground injuries.

Connections with Other Units If you are doing the full-year *Investigations* curriculum in the suggested sequence for grade 5, this is the last of nine units. Before students begin this unit, it is important that they have experiences with data analysis equivalent to those provided in the grade 4 Statistics unit, *Shape of the Data*. If they haven't, you may want to do that unit instead. The grade 4 Data and Fractions unit, *Three out of Four Like Spaghetti*, in which students first use fractions to talk about data, would be helpful preparation for *Data: Kids, Cats, and Ads*. The grade 5 Fractions, Percents, and Decimals unit, *Name That Portion*, builds skills that students will need for this unit. Be sure your students have those skills before starting *Data: Kids, Cats, and Ads*.

This unit can also be used successfully at grade 6, depending on the previous experience and needs of your students.

Investigations Curriculum ■ Suggested Grade 5 Sequence

Mathematical Thinking at Grade 5 (Introduction and Landmarks in the Number System)

Picturing Polygons (2-D Geometry)

Name That Portion (Fractions, Percents, and Decimals)

Between Never and Always (Probability)

Building on Numbers You Know (Computation and Estimation Strategies)

Measurement Benchmarks (Estimating and Measuring)

Patterns of Change (Tables and Graphs)

Containers and Cubes (3-D Geometry: Volume)

▶ *Data: Kids, and Ads* (Statistics)

Investigation 1 • Balancing Act

Class Sessions	Activities	Pacing	Ten-Minute Math
Session 1 (p. 18) STANDING ON ONE FOOT	Collecting Balancing Data Line Plots of the Data ■ Homework ■ Extension	1 hr	
Sessions 2 and 3 (p. 28) COMPARING STUDENT AND ADULT DATA	Fractional Parts of Data Sets Comparing Adults and Students Sharing Our Comparisons ■ Homework	2 hrs	The Digits Game
Session 4 (p. 34) MYSTERY BALANCERS	Discussing Disagreements About Data ■ Teacher Checkpoint: The Mystery Balancers Reasons for Our Guesses	1 hr	

Investigation 2 • Examining Cats

Class Sessions	Activities	Pacing	Ten-Minute Math
Session 1 (p. 40) A SET OF CATS	Describing the Cats ■ Teacher Checkpoint: Representing Cat Data Sharing Our Findings on Cats ■ Homework	1 hr	
Session 2 (p. 49) ASSOCIATIONS: ARE FEMALE CATS SHORTER?	Asking Questions About Cats Sharing Findings About Associations ■ Homework	1 hr	The Digits Game
Session 3 (p. 56) (Excursion)* USING THE COMPUTER TO INVESTIGATE CATS	Exploring with Database Software	1 hr	

Investigation 3 • Sampling Ourselves

Class Sessions	Activities	Pacing	Ten-Minute Math
Session 1 (p. 60) REVIEW: FRACTIONS, DECIMALS, PERCENTS	Using Data Strips Using a Calculator for Decimals Working with Class Poll Data ■ Homework	1 hr	
Sessions 2 and 3 (p. 66) SAMPLING THE CLASSROOM	Reviewing the Sample of Cats Sampling the Class Small-Group Sampling ■ Homework	2 hrs	Volume and Surface Area
Session 4 (p. 76) THE CLASSROOM AS A SAMPLE	Are We a Good Sample? Comparing the Survey Results ■ Homework	1 hr	

*Excursions can be omitted without harming the integrity or continuity of the unit, but offer good mathematical work if you have time to include them.

Investigation 4 • A Sample of Ads

Class Sessions	Activities	Pacing	Ten-Minute Math
Session 1 (p. 84) FRACTIONS OF NEWSPAPER PAGES	■ Teacher Checkpoint: How Much of the Page Is Ads? ■ Homework	1 hr	
Session 2 (p. 87) COLLECTING DATA FROM TEN DAYS	Sampling and Collecting Ad Data ■ Homework	1 hr	Volume and Surface
Session 3 (p. 91) COMBINING AD DATA ACROSS PAGES	Putting the Fractions Together Drawing Conclusions from the Sample ■ Extensions	1 hr	

Investigation 5 • Researching Play Injuries

Class Sessions	Activities	Pacing	
Session 1 (p. 98) ISSUES OF PLAYGROUND SAFETY	Considering the Problem Defining the Survey Questions ■ Homework	1 hr	
Session 2 (p. 107) COLLECTING PLAYGROUND DATA	Refining the Questions Choosing a Sample ■ Homework	1 hr	
Sessions 3, 4, and 5 (p. 110) ANALYZING AND PRESENTING DATA	Compiling the Data Making Posters of Data Analyses ■ Assessment: What We Recommend Choosing Student Work to Save ■ Homework ■ Extensions	3 hrs	

Following are the basic materials needed for the activities in this unit. Items marked with an asterisk are provided with the *Investigations* Materials Kit for grade 5.

Large clock or watches with a second hand

* Measuring tools—rulers, metersticks, or tape measures: 1 per small group. (12 metersticks are included in the kit.)

Full-color Cat Poster (optional)[1]

* Adding machine tape: 1 roll

Newspapers: 10 days' issues of the same paper, *USA Today* or your choice

Computer, database software (optional)[2]

Calculators: 1 per pair of students

Scissors: 1 per student

Glue or tape, including some that can be removed easily, for students to share

Colored markers for students to share

Stick-on notes: 1–2 small pads

3-by-5 cards or slips of paper: 6 per student

Chart paper

Construction paper (for display graphs)

Clipboards for data gathering (optional)

Overhead projector

The following materials are provided at the end of this unit as blackline masters. They are also available in classroom sets.

Family Letter (p. 126)

Student Sheets 1–12 (starting p. 127)

Teaching Resources:

Cat Cards (pp. 134–145)

Data Strips (p. 150)

Recording Strips (p. 151)

Digit Cards (p. 153)

Cube Diagrams (p. 154)

---

[1] Black-and-white photos of the cats under study are included on the blackline masters. A color poster picturing the same cats is available from Dale Seymour Publications®.

[2] Tabletop Sr.™ is an ideal tool for this unit. It is available from Dale Seymour Publications. Requirements:

Microsoft® Windows™ version: 386SX or higher; Windows 3.1; 4 MB RAM; VGA

Apple® Macintosh® version: LC, II, Centris, or equivalent; Apple System 6.0.8 or higher; 4 MB RAM; 256 color monitor

Other database software suitable for elementary students can be used but may offer fewer options.

This unit on data analysis has two intertwined emphases. The first focus is on describing, representing, and comparing data sets. In many math programs, students are doing this from the earliest grades. Unfortunately, fifth and sixth graders are typically doing the same kinds of surveys and graphs that they did when they were much younger. They may learn a few new conventions, but they are not learning how to explore data on a deeper level. This unit encourages students to delve into data and describe data sets in a more sophisticated way. This means that they

- describe data sets that involve several variables
- describe the similarities and differences between data sets—including data sets that are not the same size
- begin to see relationships between variables within a data set
- construct representations that allow them to readily compare two data sets
- use data sets as evidence for comparative statements

The unit's second focus is on sampling: What is a sample? What is it good for? What makes it fair? Sampling is a complex area of statistics, and statisticians and nonexperts alike wonder how big a sample needs to be and how to make sure that a sample reflects the population well. The main mathematical idea explored here is how to choose a sample that is reasonable. One goal is to help students make sensible decisions as they choose a sample. Another goal is to teach them how to avoid obvious sources of bias in selecting a sample. Having these experiences gives students the expertise to evaluate studies done by others.

As students work with sampling, they will make use of and expand their understanding of fractions and percentages. In order to compare what is true of a sample with what is true of a population, students will compare groups that are very different in size. This usually necessitates the use of fractions and percentages. For example, if 10 out of 25 students in your classroom help out at home by setting the table, how does your class compare to a national sample in which 57 percent of children set the table?

Working with sampling gives students an introduction to an important statistical area, and at the same time develops their understanding of ways of comparing parts to wholes. Specifically, the following processes are emphasized:

- determining the characteristics of a reasonable sample
- understanding what information different-sized samples provide
- comparing data from a sample with a target number
- using fractions and percentages to compare a smaller sample with a bigger sample (or with a population)

Mathematical Emphasis At the beginning of each investigation, the Mathematical Emphasis section tells you what is most important for students to learn about during that investigation. Many of these mathematical understandings and processes have multiple layers, and students gradually learn more and more about the ideas over many years of schooling. Individual students will begin and end the unit with different levels of knowledge and skill, but all will gain greater knowledge about how to collect, represent, describe, interpret, and compare data sets.

In the *Investigations* curriculum, mathematical vocabulary is introduced naturally during the activities. We don't ask students to learn definitions of new terms; rather, they come to understand such words as *factor* or *area* or *symmetry* by hearing them used frequently in discussion as they investigate new concepts. This approach is compatible with current theories of second-language acquisition, which emphasize the use of new vocabulary in meaningful contexts while students are actively involved with objects, pictures, and physical movement.

Listed below are some key words used in this unit that will not be new to most English speakers at this age level, but may be unfamiliar to students with limited English proficiency. You will want to spend additional time working on these words with your students who are learning English. If your students are working with a second-language teacher, you might enlist your colleague's aid in familiarizing students with these words, before and during this unit. In the classroom, look for opportunities for students to hear and use these words. Activities you can use to present the words are given in the appendix, Vocabulary Support for Second-Language Learners (p. 124).

seconds, minutes, score In the first investigation, students collect data on how long, measured in *seconds* and *minutes,* different groups of people (students their own age and adults) can stand balanced on one foot with their eyes closed. The individual pieces of data are recorded as *scores.*

compare, range Throughout the unit, students are asked to *compare* two or more data sets. One of the points of comparison is the *range* of the data—the interval from the lowest to the highest value in the data set.

injury, serious, emergency room In the final project of this unit, students gather data related to playground safety. They formulate survey questions about accidents that occur on their playground, focusing on issues like where and when the accidents happen, and the *seriousness* of the resulting *injury.*

Multicultural Extensions for All Students

Whenever possible, encourage students to share words, objects, customs, or any aspects of daily life from their own cultures and backgrounds that are relevant to the activities in this unit. For example, in Investigation 4, when students analyze the ads in newspapers, some students might bring in newspapers from their own communities. Some of these may be in different languages. Students can compare the size and the kind of ads that appear in different newspapers.

During Investigation 5, students can discuss playground games from different cultures. How much space do these games take on the playground? How many students participate? Does the game involve actions or equipment that might affect playground safety?

Investigations

Balancing Act

What Happens

Session 1: Standing on One Foot   Students collect data about how long they can stand on one foot with their eyes closed. They use line plots to represent these data and to compare the distributions for left and right feet.

Sessions 2 and 3: Comparing Student and Adult Data   Students examine data sets collected from adults balancing on one foot and compare these data with their own. They investigate whether the fifth graders or the adults are better balancers.

Session 4: Mystery Balancers   Students discuss disagreements they had in comparing adult and student balancers in Session 3. Then, in small groups, they consider different sets of mystery data: data on balancing that have been collected from gymnasts, karate students, younger children, and people over 50. Students write a comparison of these data sets and form a hypothesis about the identity of the mystery balancers.

Mathematical Emphasis

- Using line plots to represent data sets
- Comparing two data sets
- Finding medians and other fractional parts of data sets
- Making hypotheses based on comparisons of two data sets
- Making statements based on data

What to Plan Ahead of Time

Materials

- Watches with a second hand: 1 per pair. These are needed only if there is no large class clock that shows seconds. (Session 1)
- Small stick-on notes: 1–2 small pads, to provide 4–10 half-size notes per student (Sessions 1–3)
- Chart paper (Sessions 2–3)
- Overhead projector (Sessions 2–3)

Other Preparation

- If your class needs to review how to find the *median* in a data set, prepare several sample data sets on line plots to use for this discussion during Session 1. You can quickly collect classroom data on shoe size, hair length in centimeters, age at which students got their first watch, or number of buttons on clothing,

- After Session 1, prepare Student Sheet 3, Student and Adult Balancers (p. 129), by numbering the line plots and filling in the data students collected in class before you duplicate it.
- Duplicate student sheets and teaching resources (located at the end of this unit) as follows:

For Session 1

Student Sheet 1: Collecting Data on Balancing (p. 127): 1 per student

Student Sheet 2, How Long Do Adults Balance? (p. 128): 1 per student (homework)

For Sessions 2 and 3

Student Sheet 3, Student and Adult Balancers (p. 129), 1 per student and 1 transparency (but see Other Preparation, above)

Student Sheet 4, Who Are Better Balancers? (p. 130): 1 per student (homework)

For Session 4

Student Sheet 5, Mystery Balancers Data (p. 131): 1 per student

Student Sheet 6, Writing About the Mystery Balancers (p. 132): 1 per student

Standing on One Foot

What Happens

Students collect data about how long they can stand on one foot with their eyes closed. They use line plots to represent these data and to compare the distributions for left and right feet. Student work focuses on:

- collecting data by timing
- representing data on a line plot
- comparing two sets of data on line plots

Materials

- Student Sheet 1 (1 per student)
- Clock or watches with second hand
- Small stick-on notes, cut in half (2 halves per student)
- Sample data sets for review of median (as needed)
- Student Sheet 2 (1 per student, homework)
- Family letter (1 per student)

Activity

Collecting Balancing Data

Briefly introduce the unit. Explain that for the next 3–4 weeks, the class will be working with data. They will start by collecting, representing, and analyzing data of different sorts. Later they will learn about choosing a sample—a special way of getting data.

In this activity, students will be doing their own data collection. They will start by investigating the balancing abilities of different groups of people.

To start this investigation, we'll be testing how long people in our class can stand, balancing on one foot, with their eyes closed. What do you predict? How long do you think people will be able to balance? Do you think people can balance longer on their right foot or their left foot? Do you think there will be big differences among us in how long we can balance, or will we be pretty much the same?

Record students' predictions on the board and discuss reasons for the different predictions. For example, some students might believe that everyone can balance longer on the right foot than on the left, while others might think this relates to whether a person is right- or left-handed.

There are many possible ways of collecting data on balancing. Students will use one standard procedure so they can compare themselves with one another and with other balancers. Distribute Student Sheet 1, Collecting Data on Balancing (p. 127). Using yourself and a student volunteer, demonstrate the standard procedure described on that sheet.

After the demonstration, students work in pairs to collect their data. Each pair must be able to see a clock or a watch that shows seconds. Remind students that, even though they may be able to balance for longer, they will time only up to 3 minutes on each foot. Students record their time for each foot on a separate half-size stick-on note, writing their name, which foot, and the time.

You will need to decide, as a group, what to do with times over 59 seconds. Should these be recorded as seconds, or as minutes and seconds? If your class decides to record these times in seconds, you may want to devote a few minutes to reviewing the conversion process; it's a good opportunity to practice multiplying by a multiple of 10.

Even though most fifth graders have little difficulty telling time, this activity may show that a few students need additional experience. While students are collecting data, circulate to check their work. Are there students who get confused when the time goes over 1 minute? when the second hand crosses the 12?

Tell students to keep the information sheet on the procedure for the balancing test as they will need it for homework, to collect more data at home.

Activity

Line Plots of the Data

After the pairs have collected their data, students display their results on two class line plots, one for left feet and one for right feet. Draw these line plots on the board, making sure they are long enough (the range is from 0 to 180, and students will be placing their stick-on notes along the line). For a review of setting up line plots, see the **Teacher Note,** Line Plots: A Quick Way to Show the Shape of the Data (p. 23). Ask students to help you make decisions about the line plot.

We're going to make line plots to organize this information quickly. What should be the beginning and ending numbers on the line plot? Should we mark off every second up to 3 minutes? every 5 seconds? every 10 seconds?

Be sure to draw and mark the line quickly to demonstrate that line plots are a quick and simple way to represent a set of data. It's best to use round numbers for the beginning and end of a line plot, rather than the actual smallest and largest data points. Then divide the rest of the line using intervals such as 10, 20, or 25. Put one line plot directly above the other, with endpoints aligned so students can make comparisons easily. The students place their stick-on notes on the line plots (when the data points are close together, their stick-on notes will have to overlap).

When the line plots are complete, examine the results. Look for clusters of data, gaps in the data, and unusual scores or outliers. For a discussion of these terms, see the **Teacher Notes,** The Shape of the Data: Clumps, Bumps, and Holes (p. 24), and Range and Outliers (p. 25).

Ask the class to consider what single value we could use to represent the entire data set—a "typical" value. Students may suggest quantities such as the *mode* (the most frequent value) or the *midrange* (the middle of the range; for example, if the range were 2 to 92, the midrange would be 47, halfway between 2 and 92). Others may focus on the value that has the biggest clump of data around it. These are all appropriate measures, but for this lesson, turn their attention to the median as a good indication of what's typical.

Note: You are likely to have students who suggest the arithmetic mean, or as they may call it, the average. They may know how to find it with the "add-'em-all-up-and-divide-by-the-number" technique. Although this algorithm is often taught in elementary school, research has shown that it is often not understood, even by older students and adults. At this point, it is better to stay away from the mean and the confusion it may introduce. Focusing on the median and related measures is more appropriate for upper elementary students.

Reviewing the Median As needed, take some time to review the median. For background information, see the **Teacher Note,** Finding Medians and Other Fractional Parts of Data Sets (p. 26). Although the definition appears simple, medians are quite complex, and it often takes students some time to figure out how to find the median of a data set. Present the sample data sets you have collected for student practice with finding medians (as described in Other Preparation, p. 17).

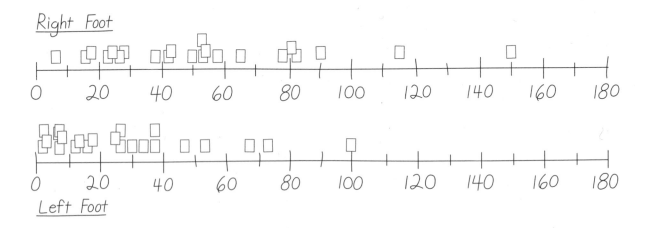

After this review, help them find the median in the balancing data they have just collected. It is helpful to talk about the relationship between a median and its data set by pointing out that half of the data set is on each side of the median, for example:

Half of our class balanced on their left leg less than 20 seconds, so 20 seconds is the median.

Describing the median in terms of one half of the data set, rather than simply stating the median as a value, leads students to think about one-fourth, one-eighth, or other fractional parts of data sets. These will be explored further in the next session.

Interpreting the Class Balancing Data   Here are some questions to guide the discussion of the two line plots:

What is "typical" of our group in terms of how long we can stand on our left foot? on our right foot? Is there much difference between people in terms of how long we can balance on each foot, or do we all look pretty similar? How can we explain any outliers or gaps in the data?

Encourage students to compare the two data sets for right-foot and left-foot balancing. If your students have not had much experience with comparing data sets, spend most of your time on this first set of questions:

What is different about the distribution of times for right feet and left feet? In general, do we balance longer on our right foot or our left? About how much longer? How did you decide? What might account for the difference?

Compare the right-foot and left-foot graphs by looking at the median of each. Use the half of the class above or below the median to compare the two distributions.

If your students are comfortable with the concept of the median, you can ask for other comparisons. They might compare the right-foot and left-foot data by looking at the fraction or percent of each graph that is above or below a particular value. For example:

What fraction of us can balance longer than 30 seconds on our right foot? on our left foot? What fraction can balance longer than 1 minute?

In this discussion, refer back to the predictions students made at the beginning of the session.

How were your actual findings different from your predictions? What do you think accounts for the differences?

Getting Ready for Session 2 After Session 1, you need to transfer the class balancing data to Student Sheet 3, Student and Adult Balancers. On your master copy, number all four line plots using the same scale and range (based on your class line plots for the student balancing data). It is important that all four line plots have the same numbering so it will be easy to compare them visually. Using X's, add students' data to the top two plots, copying from the class line plots made with stick-on notes. Students will complete the other two line plots in the next session, using the adult data they collect for homework. After preparing Student Sheet 3, make copies for each student and 1 transparency.

Session 1 Follow-Up

 **Homework**

Send home the family letter with Student Sheet 2, How Long Do Adults Balance? Make sure students also take home their copy of Student Sheet 1, Collecting Data on Balancing. Students will collect data for one or more adults, for both right and left feet. Briefly discuss how to be sure the adult data will be comparable to the data students have collected on themselves. For example, be sure to allow adults one practice first and observe the 3-minute time limit. Also, try to have adults stand on the same kind of surface as students did (carpeted or not). Encourage students to review the testing procedure (on Student Sheet 1) with adults before they start the timed trials.

After students collect their adult data, they write two or more predictions about those data. For example, based on the information they have collected, do they think adults will balance, in general, a longer or shorter time than the students in their class? Will there be a difference between left foot and right foot for adult balancers? Who will be the overall longest and shortest balancers?

 **Extension**

How Do Individuals Differ from Right Foot to Left? In order to study whether individual students balance longer on one foot than the other, students can make a list of everyone's data, with right- and left-foot scores next to another. They can use this representation to find out how many people balance longer on each foot, who had the largest difference between left and right foot, and so on.

Name	Right foot	Left foot
Rachel	52 sec.	31 sec.
Tai	90 sec.	15 sec.

Line Plots: A Quick Way to Show the Shape of the Data

A line plot is a quick way to organize numerical data. It clearly shows the range of the data, the interval from the lowest to the highest value, and how the data are distributed over that range. Line plots work especially well for numerical data with a small range, such as the number of seconds one can balance on one foot, or, as shown below, the number of raisins packed in a standard-size box (from the *Investigations* grade 4 data unit).

A line plot is especially useful as an initial organizing tool for work with a data set. Most often used as a working graph, it need not include a title, labels, or a vertical axis. It is simply a sketch showing the values of the data along a horizontal axis and X's to mark the frequency of those values.

From the line plot showing the number of raisins in 15 boxes, we can quickly see that two-thirds of the boxes have either 37 or 38 raisins. Although the range is from 30 to 38, the interval in which most data fall is from 35 to 38. The outlier, at 30, appears to be an unusual value, separated by a considerable gap from the rest of the data.

One advantage of a line plot is that each piece of data can be recorded directly on the graph as it is collected. To set up a line plot, students start with an initial guess about what the range of the data is likely to be: What should we put as the lowest number? How high should we go? Leave some room on each end so that you can lengthen the line later if the range includes lower or higher values than you expected.

By quickly sketching data in line plots on the chalkboard, you provide a model of using such graphs to get a quick, clear picture of the shape of the data.

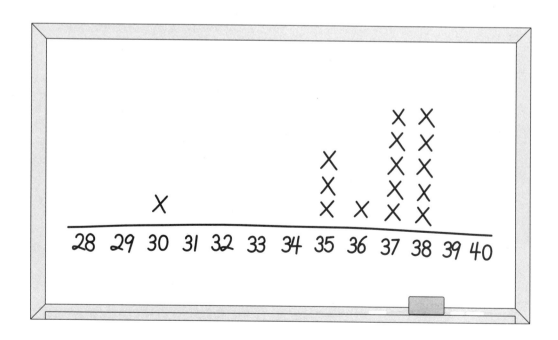

The Shape of the Data: Clumps, Bumps, and Holes

Describing and interpreting data is a skill that must be acquired. Too often, students simply read numbers or other information from a graph or table without any interpretation or understanding. It is easy for students to notice only isolated bits of information ("Vanilla got the most votes" or "Five people were 50 inches tall") without developing an overall sense of what the graph shows. To help students pay attention to the shape of the data—the patterns and special features—we have found useful such words as *clumps, clusters, bumps, gaps, holes, spread out,* and *bunched together.* Encourage students to use this casual language to describe where most of the data are, where there are no data, and where there are isolated pieces of data.

A discussion of the shape of the data often breaks down into two stages. First, we decide what are the special features of the shape: Where are the clumps or clusters, the gaps, the outliers? Are the data spread out, or are lots of the data clustered around a few values? Second, we decide how we can interpret the shape of these data: Do we have theories or experiences that might account for how the data are distributed?

As an example, consider the graph below, which shows the weight in pounds of 23 lions in U.S. zoos. In the first stage of a discussion of this

graph, one group of students observed the following special features:

■ There is a clump of lions between 400 and 475 pounds (about a third).

■ There is another cluster centering around 300 pounds (another third).

■ There are two pairs of much lighter lions, separated by a gap from the rest of the data.

In the second stage of the discussion, students considered what might account for the shape of these data. They immediately theorized that the four lightest lions must be cubs. They were, in fact, one litter of 4-month-old cubs in the Miami zoo. The other two clusters turned out to reflect the difference between the weights of adult male and female lions.

Throughout this unit, we strive to steer students away from merely reading or calculating numbers drawn from their data (the range was 23 to 48 minutes, the median was 90 days, the biggest height was 52 inches). These numbers are useful only when they are seen in the context of the overall shape and patterns of the data and when they lead to questioning and theory building. By focusing instead on the broader picture—the shape of the data—we discover what the data can tell us about the world.

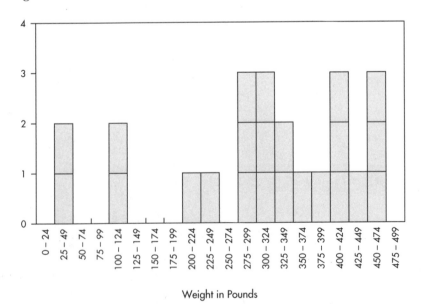

Weight in Pounds

Range and Outliers

Range and *outlier* are two statistical ideas that come up naturally in discussing data with students.

The *range* of the data is simply the interval from the lowest value to the highest value in the data set. The range of the data in the line plot below, showing the tail length of 24 cats (from Investigation 2), is from 1.5 to 13 inches.

An *outlier* is an individual piece of data that has an unusual value, much lower or much higher than most of the data. It "lies outside" the overall shape and pattern of the data. There is no one definition of how far away from the rest of the data a value must be to be termed an outlier. Although statisticians have rules of thumb for finding outliers, these are always subject to judgment about a particular data set.

As you view the shape of the data, you and your students must judge whether there are values that don't seem to fit with the rest of the data. For example, in the data on cats' tail lengths, 1.5 inches seems to be an outlier. In the line plot of raisin data (p. 23), 30 seems to be an outlier.

Both range and outlier are ideas that will come up naturally in this unit. They can be introduced as soon as they arise in the students' descriptions of their data. Students easily learn the correct terms for these ideas and are particularly interested in outliers.

Outliers should be examined closely. Sometimes they turn out to be mistakes—someone counted, measured, or recorded incorrectly. Other times they are simply unusual values. Students are generally interested in building theories about these odd values: What might account for them?

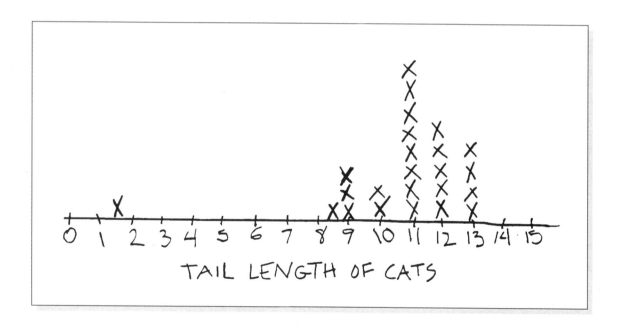

TAIL LENGTH OF CATS

Finding Medians and Other Fractional Parts of Data Sets

The median is an important landmark in a set of data. It is an average or a measure of center that helps summarize how the data are distributed. The median is the midpoint of the data set. If all the pieces of data are lined up in order, and one person counts from one end while another person counts from the other end, the value where they meet is the median value. With an odd number of pieces of data, the median is the value of the middle piece. With an even number of pieces of data, the median is the value midway between the two middle pieces.

One way to visualize the process of finding the median is to take all the values from a line plot and stretch them out in order. For example, see the line plot of student balancing times. Below it, the same data are spread out in a line, from smallest value to largest.

The middle of the student data set is 59 seconds; there are ten pieces of data above 59 seconds and ten pieces of data below it. That means 59 seconds is the median. Note that the median is not the middle of the range of the data. The range of these data is 150 (from 3 to 153), so the middle of the range would be 78 seconds (3 + 150/2), while the median is 59 seconds.

The median of the adult data set is also 59 seconds, but as the line plots show, the two distributions have very different shapes. The adults have a bunch of balancers from 2 to 19 seconds and a gap from 19 to 52, while the students' values are spread out more evenly from 3 to 50 seconds. If we look at just the medians, we would miss this big difference between the data sets, which might be of interest to us.

Student Balancing Time in Seconds

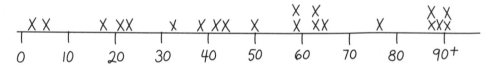

3 5 18 21 22 33 39 42 44 50 59 59 64 64 65 78 88 88 89 99 153

Adult Balancing Time in Seconds

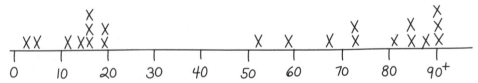

2 4 11 13 15 15 15 19 19 52 59 68 72 72 81 85 85 88 90 91 130

Continued on next page

Fractional Parts of Data Sets   Extending the idea of median gives us another tool for comparing data sets: using fractions other than one-half. For example, we can find a point that is one-fourth of the way through the data set; one-fourth of the data will be below this value, and three-fourths of the data above it. In the student data set, for example, there are 21 data points. One-fourth of 21 is about 5, so the fifth data point is about one-fourth of the way through the set. The fifth data point is 22.

If we look at the point one-fourth of the way through the adult data set (again the fifth data point), we see that it is 15. This gives us one indication that students could be considered better balancers: one-fourth of the way through their data set they are up to 22 seconds, while the adults are only up to 15 seconds.

Now let's look at the data points three-fourths of the way through these two data sets. Three-fourths of 21 is about 16, so we'll look at the sixteenth data point. For the students, that value is 78 seconds, while for the adults, it is 85 seconds. By this measure, *adults* are better balancers. This kind of seemingly contradictory result is actually common. What it means in this case is that the lower half of adult balancers are even worse than the lower half of student balancers, but on the other hand, the better half of adult balancers are even better than the better half of student balancers. Another way of describing this difference is to say that the adults' times are more clustered at the extremes of the data, while the students' are distributed more evenly.

You may have noted that to find one-fourth and three-fourths of the number of data points, we rounded rather than finding an exact quotient or finding a value between two data points, as we do with the median. In the practice of statistics, there are precise rules for finding the one-fourth, three-fourths, and other fractional values; your students may learn these later in their math careers. For now, the estimation method is appropriate for elementary students and will come very close to the results obtained with more formal methods. Most important here is that students learn the general idea of finding a value that is about *any* fraction of the way through a data set.

A related technique for comparing data sets is to observe what fraction of each data set is above or below a particular number that has special significance, such as 1 minute. In these data sets, the adults have $10/21$ balancers above a minute. The students have $9/21$ balancers above a minute. By this measure, adults are just a little better at balancing.

The data sets in this example are the same size, but, as students will see later in the unit, the fraction technique is especially useful for comparing two data sets of unequal size. In the same way as described above, we can compare data sets by finding the value one-fourth (or two-thirds, or any other fraction) of the way through each, even though this value may be the eighth value in one data set and the twenty-eighth in the other.

Comparing Student and Adult Data

Materials

- Student Sheet 3 (1 per student)
- Completed homework on Student Sheet 2
- Small stick-on notes, cut in half (2–8 halves per student)
- Chart paper
- Student Sheet 4 (1 per student, homework)
- Transparency of Student Sheet 3
- Overhead projector

What Happens

Students examine data sets collected from adults balancing on one foot and compare these data with their own. They investigate whether the fifth graders or the adults are better balancers. Student work focuses on:

- finding medians and other fractional parts of data sets
- comparing data sets
- making statements based on data

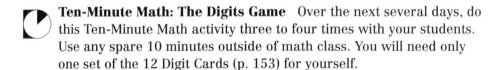

Ten-Minute Math: The Digits Game Over the next several days, do this Ten-Minute Math activity three to four times with your students. Use any spare 10 minutes outside of math class. You will need only one set of the 12 Digit Cards (p. 153) for yourself.

Choose a target number, for example, 1000. Randomly choose some Digit Cards, one for each place in the target number, plus one extra. Thus, for 1000, draw five Digit Cards. Write these five digits on the board. Using pencil and paper, students arrange the digits to make a number that is as close as possible to the target number, then compare their solutions.

For full directions and variations on this activity, see p. 120.

Activity

Fractional Parts of Data Sets

Distribute the copies of Student Sheet 3 and put the transparency on the overhead. As a whole class, examine these data to review medians and to find other fractional parts of data sets (as described in the **Teacher Note**, Finding Medians and Other Fractional Parts of Data Sets, p. 26).

These line plots are copies of our class balancing data. Yesterday we looked at the medians of our left- and right-foot data and used them to compare the two sets of data. Using the median, we have made statements like, "Half of our class balanced more than 30 seconds on their right foot."

We can make similar statements about other fractions of the data. Can you make a similar statement that uses the fraction one-quarter of our class?

If students need help, ask how many people are in the class and what one-quarter of that number would be. When they answer this question, they can look at the data set to find the value of the appropriate data point.

Can you make a similar statement that uses the fraction three-quarters? Can you make one using the fraction two-thirds? What fraction of the left-foot data is below 10 seconds? Fractions of data sets can be used to compare two different sets of data. How could we use some fraction other than one-half to compare our right-foot and left-foot balancing?

Write students' statements on the board as they suggest them. They might say something like this (varying with your actual data):

> One-quarter of the students balanced less than 10 seconds on their right foot, but one-quarter of the students balanced less than 25 seconds on their left foot.

Students also tell what their comparison leads them to conclude; for example:

> People in our class found it easier to balance on their left foot than their right.

A related kind of statement students may make, which is just as appropriate, is to choose a data value (such as 1 minute) and compare the fractional part of each data set that is above or below that number. Introduce this type of statement as well, using your class's data:

We can also say, for example, that [half] of the students balanced for 1 minute on their right leg, but only [one-third] of them balanced for 1 minute on their left leg. What would this say about how hard it was for them to balance on the right foot compared to the left foot? Can you make another statement in a similar form?

Discussing Students' Hypotheses Refer students to their completed homework on Student Sheet 2.

We'll be collecting your data on adults' balancing skills that you gathered at home, but before we do that, let's talk about your predictions. How do you think the adults' data will differ from ours? What did you write?

As students read their predictions, record them on chart paper to compare later with the results of the data analysis. Ask questions like these:

Some of you think that adults will be better able to balance, while others think that fifth graders will be better. What do you mean by "better"? How will the data look? Will *all* the data be higher for one group than for the other? Will only some of it be higher? Do you think the data will look different for the right foot and left foot? In what way?

Putting Together the Adult Data Distribute small stick-on notes, cut in half, to students, two halves for each adult they tested. Students transfer their data from Student Sheet 2 to the stick-on notes, writing on each note the length of time the adult balanced, left or right foot, and the adult's initials. Set up two more line plots on chart paper or the board, for adults' right-foot and left-foot data. Students place their stick-on notes on the line plots, just as they did during Session 1 with their own data. (If you have all four line plots on display at the same time, be sure to use the same scale on all of them to make comparisons easier to see.)

When the adult line plots are complete, students copy the data (using an X to mark each point) onto the remaining two line plots on Student Sheet 3. If students have a hard time seeing the line plots from their seats, you may want to read the data to them. Students will be using this sheet for reference in Sessions 3–4, as well as for their homework after Session 3.

Note on the Session 2 Homework Student Sheet 4, Who Are Better Balancers? (p. 130), provides homework after Session 2. Alternatively, the sheet could be done in class. This sheet presents five statements that compare adult and student balancing data for unseen data sets. It offers good practice in interpreting statements about data, as well as presenting models for the type of comparisons students will be making in the next activity. As you begin Session 3, discuss the students' responses to Student Sheet 4. Focus on the reasons they wrote for deciding whether a given statement favors adults or students. Spend extra time on the last item (creating line plots to represent one of the statements) if some students had trouble with this.

Comparing Adults and Students

When you collected your own pieces of adult data, you made some predictions about how adult balancing skills would compare with ours. Now we're going to look at all the data and see how your predictions worked out.

Refer students to their four line plots of data for student and adult balancers (Student Sheet 3), and to their list of predictions you recorded on chart paper.

In order to compare your balancing times with those of adults, we're going to look at many aspects of the data, just like statisticians do when they compare data sets. They might look at where most of the data fall, where the largest chunk of data is, what the range of the data is, or where the halfway point (median) falls. Some look at outliers—data that are far away from the rest of the group—while others might ignore outliers.

Your job is to study the data and make statements that are supported by what you see in the data. Some of your statements may seem to show that students are better balancers than adults, while others may indicate that adults are better. You will work in pairs to make a chart of different statements about the data.

On the overhead, on the board, or on a horizontal piece of chart paper, set up a model chart in six columns, like this:

Which foot?	What you are comparing	Adults	Students	Who's better?	Agree?
left	highest value	105	100	adults	

Give students time to copy your model onto lined paper. For the in-class activity, focus exclusively on either the left- or right-foot data in both sets. The other foot will be done as homework.

Model for students finding a comparison they can make between adults and students, and making the appropriate entries on the chart. In the *What you are comparing* column, write what you're reporting for each group of people, such as range, or median, or fraction above 30 seconds. Then fill in the values for adults and students. Finally, write which group seems to balance better, based on this comparison—students or adults. Explain that the last column will be filled in later.

Working in pairs, students compare the data sets in as many different ways as they can think of, using characteristics such as shape, range, clumps, high and low values, median, other fractional parts of the data sets, outliers, and so on. Remind students that some of their comparisons may support adults as better balancers, while others may support students. If your students need practice with finding medians or other fractional parts of data sets, you might require that they make one or more comparisons based on those techniques.

Sharing Our Comparisons

When student pairs have written several comparisons each, pair up the pairs to discuss their statements. Each pair shares one comparison at a time with the other two students. The pair that is sharing can put a check in the last column of the chart only if the other pair is convinced that their comparison and conclusion make sense. If the two pairs can't agree about the soundness of the comparison, the students who wrote it should write *no* or *disagree* in the last column. Pairs alternate sharing their comparisons. The **Dialogue Box,** Students Versus Adults as Balancers (p. 33), illustrates such a discussion between two pairs.

Sessions 2 and 3 Follow-Up

Homework

■ After Session 2, students do Student Sheet 4, Who Are Better Balancers? Refer to the in-text discussion of this sheet (p. 30).

❖ **Tip for the Linguistically Diverse Classroom** As a comprehension aid for students with limited English proficiency, write the words *adult* and *fifth grader* on the board, drawing distinctly taller and shorter stick figures as cues for each term. Students then sketch the appropriate stick figures wherever the words appear in the statements. Read aloud each statement on Student Sheet 4 before students leave class.

■ After Session 3, students continue the comparison chart they started in class, making two to four comparisons between adults and students on the foot they did not compare in class. Remind them to take home Student Sheet 3 along with their chart so they will have the data they need.

Students Versus Adults as Balancers

Toshi and Becky are working together with another pair of students, Marcus and Manuel, discussing their charts that compare adult and student balancers (Sharing Our Comparisons, p. 00).

Toshi: The highest adult balanced a little over 125 seconds, and the highest kids could balance more than 180 seconds. They would have gone longer, but we had to stop timing. It's obvious who won that one.

Becky: Yeah, and there are 5 kids—that's one quarter of our group—that balanced over 125 seconds.

Marcus: Okay, but there were more kids than adults!

Becky: That's why we used fractions. And besides, there was only one out of 18 adults that scored over 125 seconds, and that's a pretty low fraction. We didn't figure it out, but I know it's less than a fourth.

Marcus: Kids win on that one, but we found there were more kids who didn't do a good job balancing.

Becky: What do you mean?

Manuel: Well, only three of the 18 grown-ups—I think that's a sixth—balanced less than 20 seconds. Closer to one-quarter of the kids balanced less than 20 seconds.

Marcus: So we say that the adults came out better on that one.

Toshi: But I don't think that's fair because those kids were laughing and bothering each other, remember?

Manuel: But it's what we found! Maybe the adults were laughing when they did it too.

Toshi: Okay, you win; the adults did better on the low end. But if the kids could do it over, who knows?

Becky: Well, if you look at the middle, the kids did better. Our median was 60, and the adults' was only 45.

Toshi: It seems like our typical score was better.

Manuel: What do you mean by "typical"?

Toshi: The score in the middle. The typical kid did better, and kids also did better on the high end.

Marcus: But remember, kids aren't only the best balancers. They're also the worst!

Mystery Balancers

Materials

- Students' comparison charts from Sessions 2–3, for discussion
- Student Sheet 5 (1 per student)
- Student Sheet 6 (1 per student)

What Happens

Students discuss disagreements they had in comparing adult and student balancers in Session 3. Then, in small groups, they consider different sets of mystery data: data on balancing that have been collected from gymnasts, karate students, younger children, and people over 50. Students write a comparison of these data sets and form a hypothesis about the identity of the mystery balancers. Their work focuses on:

- choosing aspects of data sets to compare
- using characteristics of data to identify unlabeled data sets

---

Discussing Disagreements About Data

First discuss the homework, asking students to share statements from their charts that compare student and adult balancers. Focus on the kinds of evidence students used. While students are looking at their charts, call attention to the last column. Ask them to describe some of the disagreements that arose when they discussed their comparisons with another pair in the previous session. For this discussion, a student pair presents their original statement; then the pair who disagreed presents their argument. The rest of the class offers their comments.

One point likely to come up in these discussions is that there are unequal numbers of adults and students (as seen in the **Dialogue Box,** Students Versus Adults as Balancers, p. 33). Some students may feel that it is impossible to compare the groups because of their different sizes. They may make remarks like, "There's more adults than kids... If you had the same number of adults as kids, there wouldn't be as many adults that were that low."

These students are struggling with one of the central concepts of data analysis: sample size. They have an intuition about the size of the sample that is right on target: Larger samples allow us to estimate characteristics of a population more accurately. (There is more on this topic in Investigation 3.) Nonetheless, it *is* possible to compare two samples of different sizes by looking at fractions (or decimals or percents) of each sample.

On the other hand, many students try to compare by counting individuals rather than by calculating ratios. These students may say that if there are three students with times under 10 seconds and four adults with times under 10 seconds, students are better balancers, regardless of what *proportion* of the sample each group represents. Watch for this kind of misunderstanding about comparing samples and clarify during the discussion.

Keep in mind that the purpose of these activities is to focus on the process of making statements that are based on data, not to determine who were the best balancers. It is likely that there will be data to support both sides.

Teacher Checkpoint

The Mystery Balancers

Distribute Student Sheet 5, Mystery Balancers Data, and Student Sheet 6, Writing About the Mystery Balancers. Students will also need their copy of Student Sheet 3 or your large class line plots of the data they collected about themselves. Explain that four additional data sets have been collected on balancing on one foot, following the same test procedures that students used themselves. One data set is from young gymnasts, one from people over 50, one from first and second graders, and one from karate students.

❖ **Tip for the Linguistically Diverse Classroom** When introducing the mystery groups, draw attention to the pictures illustrating them on Student Sheet 6.

Group Work Divide the class into groups of three or four, and assign each small group one mystery data set. (More than one group may work with the same data.)

In their groups, students spend a short time comparing the data on themselves with that of the mystery balancers, and make a guess about which group is represented by the data they have been assigned. They write their guess (for example, "Data set B is gymnasts") on a sheet of paper and give it to you, to be compared with the actual answers later.

Individual Work After groups have considered the mystery data, each student writes a brief report comparing the set of mystery balancers to the data on their class. The questions on Student Sheet 6 are guidelines for this report. Make sure students are referring to both their own student data and the mystery data as they write their responses.

Circulate to observe how students are working with the data. When everyone has finished, collect the papers. Here are some questions to consider as you review their work:

- Do they support their opinions with data?
- Do they go beyond the simplest descriptions of highest and lowest values?
- Do they use any characteristics of the data that require using fractions or fractional parts of data sets, such as median?
- Do they combine several individual statements to make a more complex and convincing statement? The **Dialogue Box,** Students Versus Adults as Balancers (p. 33), illustrates some examples of complex statements comparing two data sets.

Activity

Reasons for Our Guesses

After group work in the previous activity, you collected each group's guess at the identity of their mystery balancers. Now share these guesses with the class and ask each group for a brief explanation of why they made their guess. Then dramatically reveal the actual identities of the mystery groups:

A—gymnasts

B—first and second graders

C—karate students

D—people over 50

Talk about why these groups' data might look the way they do. Some of the groups have similar data patterns (for example, people over 50 and the first and second graders), so there may be no way to distinguish between them based on the information here.

Emphasize that the important part of this exercise is not guessing the right group, but being able to give reasons based on the data for why a particular line plot might correspond to a certain group. Remind students that they will be using this kind of reasoning for the rest of this unit.

After students have described the reasons for their choices, share the following information that helps explain the differences among the groups.

Were you surprised to find out who the mystery balancers were? Here is some information that might help you make sense of the data.

- While common sense tells us that gymnasts should be great balancers, gymnasts are in fact similar to other students their age. Gymnasts carefully use their eyes to focus on particular spots, and they learn just where to look when they are balancing. When their eyes are closed, as they were for our balancing test, they no longer have this advantage.

- First and second graders are still learning balance skills. They are practicing all the time, through games and other activities. But they haven't yet mastered balancing the way that older children have.

- The karate students tested had been involved in their sport between 1 and 11 years. They were serious about completing the work needed to achieve higher levels of "belts." Karate involves considerable skill at balancing, as is reflected in their scores.

- As people age, many of their senses become less acute, and that includes the sense of balance. A group that is over 50 could include people who are quite elderly and whose balance is thus not good.

Would any of you have guessed differently if you had this information earlier? If so, what information would have influenced you?

Examining Cats

What Happens

Session 1: A Set of Cats Students receive a set of Cat Cards, each of which provides data on one cat: weight, length, age, and gender as well as eye, pad, and fur color. In small groups, students focus on one particular variable as they make representations describing this group of cats.

Session 2: Associations: Are Female Cats Shorter? Students consider associations between two variables, such as length and gender, and determine how two variables might "go together."

Session 3 (Excursion): Using the Computer to Investigate Cats Students add their own cat data to a computer database and use it to explore questions about the relationships between variables.

Mathematical Emphasis

- Collecting and examining data that involve more than one variable

- Making representations of numerical and categorical variables

- Using fractions and percentages to understand categorical and numerical data

- Exploring the way two variables in a data set might be related

- Learning to enter and analyze data in a computer database

What to Plan Ahead of Time

Materials

- Rulers or metersticks: several for reference (Session 1)

- Full-color Cat Poster (available from Dale Seymour Publications), to supplement the reproducible cards with black-and-white photos (Sessions 1–2, optional)

- Materials for making representations of data: chart paper, construction paper, scissors, colored markers (Sessions 1–2)

- Stick-on notes: 1–3 per student (Session 2)

- Database software, such as Tabletop Sr. (Excursion, Session 3)

- Computer: at least 1 (Excursion, Session 3)

Other Preparation

- Duplicate student sheets and teaching resources (located at the end of this unit) as follows:

For Session 1

Cat Cards (pp. 134–145): 1 set per 2–3 students. If possible, copy them onto card stock for durability.

Student Sheet 7, Collecting Cat Data (p. 133): several per student (optional, homework)

A Set of Cats

Materials

- Rulers or metersticks (several, for reference)
- Cat Cards (1 set per 2–3 students)
- Full-color Cat Cards or Cat Poster (optional)
- Chart paper, construction paper, scissors, colored markers
- Student Sheet 7 (several per student, optional, homework)

What Happens

Students receive a set of Cat Cards, each of which provides data on one cat: weight, length, age, and gender as well as eye, pad, and fur color. In small groups, students focus on one particular variable as they make representations describing this group of cats. Student work focuses on:

- making representations of numerical and categorical variables
- using data representations to describe numerical and categorical variables

Activity

Describing the Cats

Begin this session by telling students to listen as you read the scientific description of a familiar animal (from R. M. Nowak and J. L. Paradiso, *Walker's Mammals of the World,* Vol. 2, 4th ed. [Baltimore: The Johns Hopkins University Press, 1983], p. 1068):

"There are more than 30 different [domestic] breeds ... The average measurements of several popular breeds are: head and body length, 460 mm [18.11 inches], and tail length, 300 mm [11.81 inches]."

What animal do you think is being described here?

Students discuss what animal this might be, given the measurements. They may want to use rulers or metersticks to help them visualize the lengths in the description. Whether or not they guess, explain that these data are about the common house cat.

What other things might they have said to describe an average cat? How do you think the scientists got the data that they used for these average measurements?

We're going to do our own study of cats in this investigation to see what we could add to the description of the average or typical cat.

Assign students to groups of two or three and hand out the Cat Cards, one packet for each group. There should be at least seven groups, so that each of the seven characteristics is studied by at least one group. Give students time to browse through the cards. Allow them to make observations or ask questions about the data set. If you have the set of full-color Cat Cards or the Cat Poster, find a way to display them for student reference.

Note: This sample of cats was not chosen to be representative; in fact, it is a classic example of a convenience sample. Most of the cats were pets of *Investigations* writers, their coworkers, and friends. A few were photographed at a cat show whose entrants included pets and shelter animals awaiting adoption. You may want to discuss this when sampling is introduced in Investigation 3 (see p. 66). Students might consider then what would make this sample more representative. For now, it is enough for students to know that these are all ordinary house cats like ones they may know.

You are going to describe and summarize the cat data on these cards. Then we can discover what is typical and what is not so typical of this group of cats. On the cards, you'll find data for each cat. You may be able to find out some other things about the cats from the pictures.

Each group will be doing research on *one* characteristic of the cats. If you finish with one characteristic, you may choose another one to study and report on.

Students may not be sure what you mean by "describe and summarize the cat data." Recall how they looked at the data on foot balancing and described it: They noticed how spread out the data were, how the data clumped together at different points, and any unusually high or low values in the data. Sometimes they talked about how much of the data were above or below a given value, saying, for example, that half of the data were located in a certain range, or that the bottom quarter of the data were below a particular value. These were all ways of describing and summarizing data.

In the cat data, some characteristics involve numbers, such as weights and ages, but others do not—details such as eye color and fur color. These non-numerical data are sometimes called "categorical" characteristics, because they are grouped in categories rather than arranged according to number.

In the cat database, which characteristics are categorical?

Point out to students that with categorical data (eye color, paw pad color, and fur color), they may want to look at the fraction of cats that falls in each category. Also point out that some of the categories are hard to define.

Some of these cats have fur of several colors, so you can't categorize each cat by a single color, such as *black* or *white*. If your group is looking at fur color (or pad or eye color), you will have to figure out what to do when a cat has more than one color.

Your group needs to report back to us in about 20 minutes to describe and summarize what you found out about the cats. Bring a chart, a graph, or some other representation to illustrate your findings. Write at least two sentences describing the characteristic you studied. Be sure you say what is typical of the cats, and how some cats differ from the typical cat. We will then make a list of what you can say is typical of cats for the characteristic you studied.

Assign one characteristic to each group. If you let students choose, be sure each characteristic is being studied by at least one group.

As soon as they get the Cat Cards, some students may want to investigate relationships that involve two or more variables in the cat data set—such as the relation between age and weight, or between gender and length. If you want to be sure that students focus on one variable at a time as a review of the process of data representation and analysis, ask them to save their investigation of two or more variables for the next session.

Activity

Teacher Checkpoint

Representing Cat Data

Make available construction paper, chart paper, scissors, and colored markers for students' representations. Give the groups 20 minutes to work. Remind them that they are to make several observations about their data as well as making a representation.

As the groups work, circulate to observe how they are examining the data. Here are some things to look for:

- Do students consider the data set as a whole, or are they focusing on just a few cats?
- To what extent are students able to summarize all the data in a succinct and accurate way?
- Do students make use of fractions rather than simply listing the numbers? For example, "More than half of the cats are gray" or "Two-thirds of the cats weigh between 8 and 12 pounds."

- Are students able to make representations that summarize and reflect the data?

- Do students understand how the data vary? Do they understand what is typical of the data?

Refer to the **Teacher Note,** Taking Information from the Cats Database (p. 45), for more detail about the cat data and for questions you might raise as you watch students work.

Some students get involved in making their representations artistic, coloring the graphs elaborately or drawing individual cats for each data point. While a certain amount of art can enhance graphs and motivate students, be sure students are not neglecting the mathematical aspects of the task in favor of the artistic ones. Remind them that the purpose of making graphs is to communicate mathematical relationships, not to create artistic products. As the unit proceeds, you may have to repeat this reminder. Refer to the **Teacher Note,** Commenting on Students' Representations (p. 46), as you look at students' graphs.

Activity

Sharing Our Findings on Cats

After 20 minutes, ask small groups of students to report to the class one or two major findings about the characteristic they examined. They should describe both what is typical of the cats and how the cats vary with respect to that characteristic. Keep track of these findings on the board or a large sheet of chart paper. Some possible major findings are shown below.

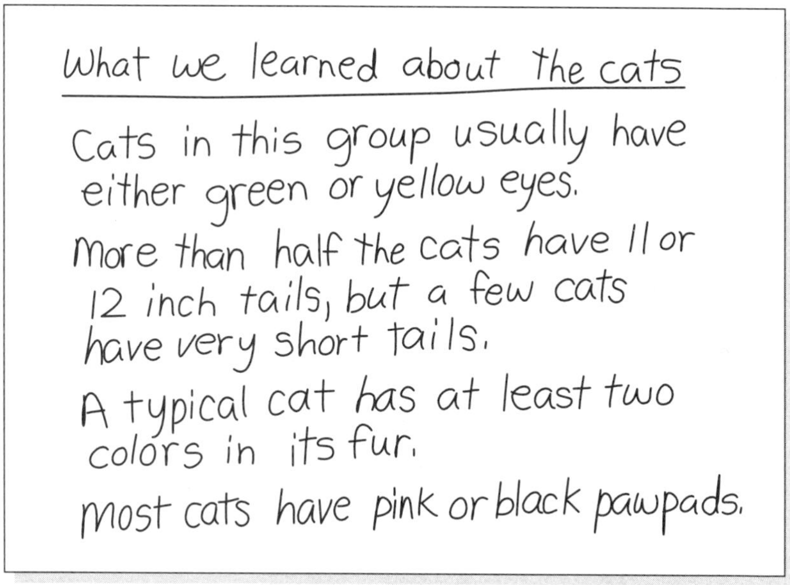

What we learned about the cats

Cats in this group usually have either green or yellow eyes.

More than half the cats have 11 or 12 inch tails, but a few cats have very short tails.

A typical cat has at least two colors in its fur.

Most cats have pink or black pawpads.

Session 1 Follow-Up

🏠 **Homework**

■ This homework—gathering more cat data—is optional. If students do it, their work can be used in the Excursion following Session 2. Even if you choose not to do the Excursion, students may still benefit from doing the homework and adding their own data to the Cat Cards for their analyses in Session 2.

In order to get a bigger sample of cats, you will collect some more cat data for homework. You will need a cat to do this homework. If you don't have a cat, ask whether you can collect data from a neighbor's or a friend's cat. Remember, though, that you have to be careful. Please don't just walk up to a strange cat and try to measure it. Ask permission from the owner and make sure you have someone to help you. If you are allergic to cats or afraid of them, see if you can find someone who will do the measurements with you, while you write down the data.

Hand out copies of Student Sheet 7, Collecting Cat Data. Explain that the section "Other" is for general comments, where they can write anything about the cat that interests them. Refer to the Cat Cards for examples.

❖ **Tip for the Linguistically Diverse Classroom**   Read aloud each heading on the chart, relating them to pictures of the cats as needed. Students make a quick drawing on their chart as a visual reminder of each category. For example, beside the words *Eye color,* they might draw two eyes, one green and one yellow.

Ask students how they will approach the task. Role-playing the activity of measuring a cat may help prepare your students for the job. See the **Teacher Note,** Measuring a Cat (p. 48), for further information.

If students like, they can make their own cat cards similar to the ones that are included with the unit, attaching snapshots or drawings to their cards. If you plan to do the Excursion, keep the set of student-made Cat Cards where students can examine them when they have free time.

■ If your students will not be collecting data on cats for homework, you might give students a digits problem from the Ten-Minute Math Digits Game (see p. 120). Select five digits and ask students to write them down. For example:

Use the digits 0, 7, 8, 3, 1 to make a number as close as possible to 1000.

Then make a number as close as possible to 500, using the same digits. Make a number as close as possible to 100. Make a number as close as possible to 10,000.

The next day, compare answers around the class.

Taking Information from the Cats Database

Students work with the Cat Cards (and, if you do the Excursion, with a computer database) to sort and organize information in order to come up with a description of a typical cat. What could students reasonably conclude from the data on the Cat Cards?

Color The most important idea about color is that there is wide variation, both in color and in pattern. However, there are some limits to this variation. Questions on this subject might include: What appear to be the most common colors of cats? How many fur colors are there? Are many cats striped?

Age As one student pointed out, "Even though there's no one-month-old cat here, we know they all were one month old once." We can tell from this sample that cats can live to be 18 years old. Questions: How long does a typical cat live? Is 18 a typical age for a cat?

Weight These data are very spread out, ranging from 6.5 to 18 pounds. There is a large clump of cats—half the sample—from 9 to 12 pounds. If we look at cats weighting 8 to 12 pounds, we account for nearly two-thirds of the cats. As more cats are added to the data, keep an eye on weight to see whether it changes. Are the younger cats lighter than the older cats?

Body Length The cats' body lengths range from 14 to 24 inches, with a median of 18.5 inches. There are two strong modes, one at 17 inches and one at 21 inches. Three-fourths of the cats fall between these two modes. In this data set, the median seems to be a good indicator of where the data are centered. Questions: Are the shortest cats younger? Are the longest cats also the heaviest cats? As more data come in, does it still appear to be true that cats generally range from about 17 to 21 inches? Are the two longest and the two shortest really unusual in the cat population?

Tail Length These data are very compact. More than half of the cats have either 11- or 12-inch tails. Adding the 13-inch tails accounts for two-thirds of the data. The median tail length is 11 inches. It could be argued that the typical cat has an 11- or 12-inch tail, and that the range is from 1.5 to 13 inches. Although the typical cat's tail may be in the 11- to 12-inch range, some cats do have very short tails. Questions: Is tail length related to body length? As data come in from your class, does the typical tail length change, or does this clump hold true?

Eye Color Cats seem, from these data, to have green or yellow eyes, with very little difference in the amount of data at each of the two values. Questions: Can cats have any other eye color? Do certain eye colors go with certain fur or pad colors? As data are added, do the yellow and green eyes continue to be represented evenly in the larger sample?

Pad Color Pad color data have two modes. Most cats have either pink or black pads. However, from this sample we can see other colors, and we can also see that cats' pads may have mixed colors. Questions: Is pad color related to fur color? Are any new colors added as your data come in?

Other It's important to remind your students that other information about cats is helpful and interesting; for example, it is interesting to note that Wally and Peebles are brother and sister. If your students add their own data, encourage them to include such comments. Although this information cannot be analyzed statistically, it can add a lot to the database.

Commenting on Students' Representations

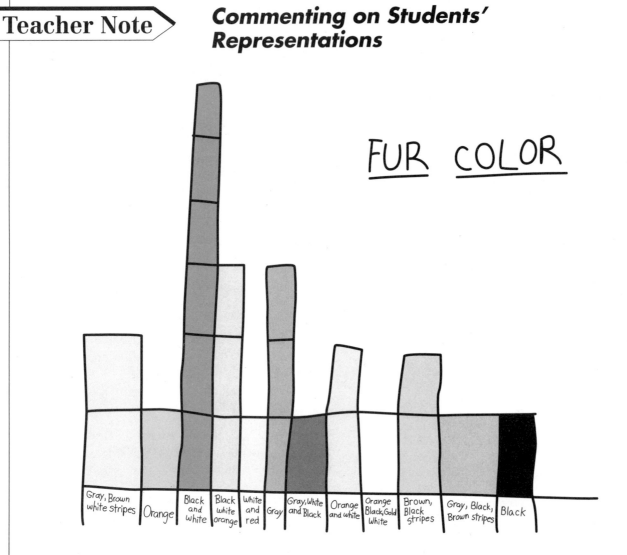

FUR COLOR

Gray, Brown white stripes | Orange | Black and white | Black white orange | White and red | Gray | Gray, White and Black | Orange and white | Orange Black,Gold White | Brown, Black stripes | Gray, Black, Brown stripes | Black

The graphs shown on these two pages are typical of what fifth grade students construct for the activity Representing Cat Data (p. 42).

Fur Color Bar Graph (above)   This representation of fur color is correct as far as it goes, but it is hard to interpret. More useful would be some grouping of the fur colors into larger categories, such as "orange and orange with something else," "black and white with more," and so on. Another possible set of categories might be "one color," "two colors," and "three or more colors." If the cats were categorized in this way, it would be easier to make observations about the typical cat, for example, "The typical cat has two or more colors in its fur."

Cat Eye Color (below)   This graph illustrates an interesting way to deal with categorical data that the students gathered outside of class. It shows three cats with green eyes, four with gold, two with yellow, and two with greenish-gold placed between the green and the gold. It is a good way to begin to make statements about eye color in a set of cats because it shows relationships between categories.

Eye Color

		X	
X		X	
X	X	X	X
X	X	X	X

green   gold   yellow

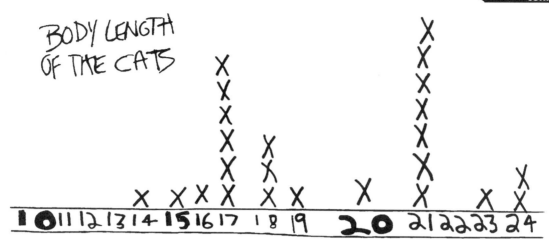

BODY LENGTH OF THE CATS

10 11 12 13 14 15 16 17 18 19 20 21 22 23 24

Cat Body Lengths (above) This graph is simple but clear. The x-axis is marked in inches, and each cat's body length is indicated by an X. This display makes it easy to observe that most of the cats (over 50 percent) are either 17 or 21 inches long, that the shortest is 14 inches and the longest 24 inches, and that the median is between 18 and 19 inches.

Cat Tail Lengths (below) This representation is artistically clever, but difficult to understand. The illustrations detract from the numbers, and the information is not arranged in a way that allows a reader to see the shape of the data.

1.5 Tomodachi Joto	12 in. Ravena
8.5 Gray kitty	12 in. Weary
9 in Pepper	12 in Diva
9 in Cleopatra	12 in Wally can't see tail
9 in Odd Fuzz / can't see his tail in picture	12 in K.C.
10 in Tigger →	13 in. Charcoal
11 in Lady Jane Gray	13 in. Lady
11 in Alexander	13 in Peau - De - Soie
11 in Melissa	13 in George
11 in. Harmony	
11 in. Peebles	
11 in Misty	
11 in. Mittens	
11 in. Strawberry	
11 in. Augustus	

If students have tried to collect their own cat data, you may hear about measurement problems.

> I couldn't weigh my cat because she wouldn't stand on the scale.

> How come Kim got 34 inches for her cat's length, and my cat and all the others we saw only got around 19 or 20 inches?

Students should know that measurement methods, agreement on the landmarks of the object being measured, and efforts at standardizing measurement are important parts of all statistics research. These processes allow mathematicians to collect data that are consistent. Measuring live animals, of course, presents some unique challenges.

It's a good idea to talk with your students about the measurements they will be collecting so that they can agree on starting and ending points for various parts, such as the tail. Such a discussion is most effective when students can think about the problem and make their own decisions about the "marking-off places" on the cats.

Students will also benefit from thinking in advance about the necessary equipment and the help they will need to stretch out a cat's tail and to get a cat to hold still. As we measured our own cat sample, we became quite skilled at rolling rulers or tape measures from cats' noses to their tail bases.

One teacher brought a stuffed toy cat to class and had students demonstrate on it how they would get the measurements from a real cat. Some students used measuring tapes and some tried with rulers and yardsticks. The students had a good laugh, and they were able to wrestle in advance with some of the technical problems they were about to face.

Weighing a cat can be a difficult matter. Owners may be able to tell students how much a cat weighs from its veterinary records. Some students will suggest that they weigh themselves with and without the cat and subtract the difference. In a home where a baby scale is available, students can weigh the cat directly—if the cat is willing.

Encourage students to enlist a helper for the measuring process. Some cats are so wiggly that it's hard to get any kind of measurement without someone else holding them still.

Associations: Are Female Cats Shorter?

What Happens

Students consider associations between two variables, such as length and gender, and determine how two variables might "go together." Their work focuses on:

- framing questions about associations between variables
- constructing representations and descriptions that help answer questions about associations between variables

 Ten-Minute Math: The Digits Game During the next few days, continue to find time for this activity outside the math hour. You might try some of the variations.

For example, present a set of four randomly drawn digits and ask students to make the largest and smallest number they can using those digits. They could use the same set of digits to create a two-part addition problem, making the largest and smallest sum they can.

See p. 120 for further discussion of the variations.

Materials

- Cat Cards (1 set per 2–3 students)
- Full-color Cat Cards or Cat Poster (optional)
- Chart paper, construction paper, scissors, colored markers
- Stick-on notes (1–3 per student)

Activity

Before You Begin If students have gathered additional cat data for homework, make copies of those data for each group so everyone will be working with the same data set.

Looking for Related Characteristics For this activity, students work in the same small groups as last session. Each group again needs a set of Cat Cards and materials for making representations.

While you were exploring the cat data, some questions came up about cat characteristics that "go together." *[Use examples you observed students thinking about, or offer the following.]* **One question is whether pad color relates to fur color: Do cats with lots of fur colors have at least two colors on their paw pads?**

Asking Questions About Cats

Another question is about how age and weight go together. Are the older cats heavier? Or are they lighter because they lose weight when they get very old? You can often find interesting information in a data set by looking at these kinds of relationships.

Before students break into small groups, brainstorm some questions about relationships between two characteristics that are of interest to them. This shouldn't just be a list of all possible relationships, but rather a thoughtful list of questions about associations that have arisen from students' exploration of the database. Students should have a hunch or hypothesis in mind as they raise the questions. Here are some examples:

- Several students know that calico cats are most often females. They ask, "Are all cats with three or more colors in their fur females?"

- Students recognize that in humans, females are generally shorter than males. They ask, "Are male cats longer than female cats?" and "Are male cats heavier than female cats?"

- Seeing that cats vary in body length and tail length, students ask, "Do body length and tail length go together? Do longer cats have longer tails?"

Students probably have several questions that have emerged from their work in Session 1. Help them shape their questions as needed, and help them identify questions that can be answered by looking at the data.

As a whole class, ask groups for the questions they thought of. List them on the board or on chart paper. Once you have several questions listed, choose one and ask students for ideas about how to investigate it, as illustrated in the **Dialogue Box**, Relationships in the Cat Data (p. 54).

Now you'll work in small groups and choose one question to explore. It can be one that we've listed or a different one that your group thinks of. Write your question at the top of a piece of paper, then decide how you're going to organize the data so you can answer the question.

You'll probably need to try a couple of ways of organizing the data before you find one that you really like. You might try a chart or a graph or writing the data in a list. Once you've organized the data, see how you can use the data to answer the question you've written down.

Students spend the next 20 minutes or so choosing a question, working with the cards, and making a representation. Just as in the last session, the teacher's role is to circulate from group to group and listen to students as they formulate questions, organize data, and attempt to answer the questions. See the **Teacher Note**, Ways of Exploring and Representing Associations (p. 52), for ways students may show associations and how you might coach them to come up with appropriate representations.

Sharing Findings About Associations

As students finish their work, they post their representations. If two or more groups have addressed the same question, they post their work next to each other. Allow students some time to examine the different representations. Provide stick-on notes so that if students have a question or observation about a specific representation, they can write it on the note and stick it to the representation. These questions will be part of the discussion to follow. Then bring the class together to hear each group give a brief description of its findings. Write on the board the points each group should cover:

1. your question
2. how you organized the data
3. how you used the data to answer the question

As each group presents their findings, you and the other students should feel free to ask questions about the relationships they found. Include the questions and observations students posted on the representations. If two groups investigated the same question, did they come to the same conclusions? If not, why not?

Keep these representations posted for the next few days and invite students to look at them during their free time.

Session 2 Follow-Up

Homework

Give students another digits problem from the Ten-Minute Math Digits Game variations. For example:

Use the following digits to make an addition problem, using two 2-digit numbers, with a sum as close as possible to 100: 7, 3, 5, 2.

Use the same digits to make a problem, using one 3-digit and one 1-digit number, with a sum as close as possible to 500. Make another such problem with a sum as close as possible to 1000.

The next day, compare answers around the class.

Ways of Exploring and Representing Associations

Exploring associations between two or more measured quantities is one of the central topics of statistics. There are many techniques for doing this formally, including correlation. In this unit, students will start to explore associations in data. They are certainly not ready for all the complexities of the topic, but there are many routes into this kind of analysis that are accessible to fifth graders. For now, most students will approach the association task by sorting (the Cat Cards make this easy), making lists of data, and making simple graphs.

One of the most common techniques students will use is to divide the Cat Cards into piles based on one categorical variable and then to construct two or more simple graphs for the other variable. For example, two students working on the association between gender and fur color first separated the Cat Cards into male and female piles. They made the following list, and then a pair of bar graphs.

From the graphs, the students drew the following conclusions:

> Only one female cat has black and white fur. But the most popular color for males is black and white fur.

> Most males (9 out of 12) have one or two colors in their fur. Over half the females have three or more colors in their fur.

In this case, the students' table and graphs were equally useful in examining the association between gender and fur color. Their data were organized in such a way that it was easy to describe the association.

Students beginning to explore association may make mistakes in organizing and interpreting their data. For example, in one class, a group of students wanted to see if there was a relationship between fur color and eye color. One student suggested the following data organization:

Cat	Eye color	Fur color	Fur color = Eye color
Gray Kitty	green	gray	no
Charcoal	yellow	bl & wh	no

Only when they finished their list and saw that the last column was made up entirely of no's did these students realize that a yes in the last column would occur only if the cat's eye and fur color matched—a very unlikely event!

A better approach would have been to put all of the yellow-eyed cats in one list by fur color and all the green-eyed cats in another list by fur color, like this:

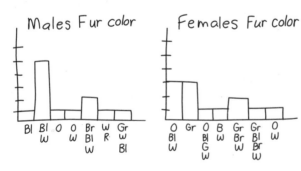

Male fur color

black	1
black & white	5
orange	1
orange + white	1
brown + black + white	2
white + red	1
gray + white + black	1

most black + white 5

Female fur color

orange + black + white	3
gray	3
orange black gold white	1
black + white	1
gray + brown + white	2
gray black brown white	1
orange + white	1

Yellow Eyes
Oddfuzz - orange, white
Melissa - white, black, orange

Green Eyes
Gray Kitty - gray
Diva - gray, black, brown, white

Continued on next page

After completing such a list, students could group the fur colors in each column to make a pair of bar charts and do further analysis. A question they might try to answer is, "Do cats with yellow eyes more often have white in their fur than cats with green eyes?"

Looking for an association between a categorical variable such as gender and a numerical variable such as body length requires students to make line plots and to compare them as they did in Investigation 1. For example, investigating the association between gender and length can be done by comparing the distribution of length for males with that for females. As in the examples above, students may first divide the Cat Cards into piles, one for males and one for females. The next step might be making two line plots, as below.

Males
14 15 16 17 18 19 20 21 22 23 24 25

Females
14 15 16 17 18 19 20 21 22 23 24 25

Then the students could compare the two graphs, noting, for example, that more than half the males are longer than 20 inches, while only one-quarter of the females are that long. They could say, then, that male cats are likely to be longer than female cats. In associating variables, the final representation is often a set of graphs, one for each value of the categorical variable. In the case of gender, there are two graphs. If we were looking at the association between pad

color and body length, there would be more graphs—a distribution of body length for each color of pad.

Some students may combine the two body length graphs into one graph, with different symbols for male and female cats, as below. This is also an excellent representation, making it relatively easy to talk about differences between males and females.

14 15 16 17 18 19 20 21 22 23 24 25

The most difficult associations for students to work with will be those between two numerical variables, such as body length and tail length. In more advanced statistics, this could be done with a scatter plot and a correlation coefficient. One way for students to approach such a relationship would be to define categories for at least one of the variables, such as "long," "short," and "medium," deciding what range of lengths would fit in each category. Then they could make a list for each category, using the techniques described above. See the **Dialogue Box,** Relationships in the Cat Data (p. 54), for an example of students exploring the association between two numerical variables.

Relationships in the Cat Data

In Session 2 of the investigation of a sample of cats, the students begin asking questions about certain characteristics that might be associated. This class has brainstormed some questions. Now they are talking about how they could explore such issues.

Name	Age	Weight
Peebles	5	9
melissa	8	11
Cleopatra	4	7
Strawberry	16	14.5
Augustus	2	10

Let's take one question to think about as a group. Which one would you like to start with?

Kevin: Are older cats heavier?

How could we organize the data so we'd be able to work with that question?

Desiree: Just make a list of all the ages. Then make a list of all the weights.

Rachel: But you need to know what things go together. It can't just be any old list.

Say a little more about that.

Rachel: Well, I'd list Cat 1 is Peebles, she's 5 years old and weighs 9 pounds. So you keep her age and weight together.

Why is that important?

Noah: Because you're thinking about whether they go together or not, so you need to keep the data from each cat together.

[The teacher writes three columns on board: Name, Age, and Weight, and begins to make a list, starting with Peebles.]

I'm writing down the name of each cat so we can keep track of who we've listed.

[Students start reading off the data; the teacher writes down the information in the appropriate column. Once they have a few rows, the dialogue continues.]

We have a few cats listed now, but how are we going to tell whether there's a relationship between the age of all the cats and their weights?

Alani: I know! Let's make a group of all the kittens and see how much they weigh, and then a group of all the grown-ups and see how much they weigh.

How will you define a kitten and a grown-up?

Alani: I think a kitten is anything less than 2 years.

Noah: That won't work, because we only have a couple of cats that young. We won't be able to tell anything.

What would you suggest then?

Noah: I'd make a list in order, maybe start with the young cats, and list them all the way up to the old cats.

Which cat should I list first?

Lindsay: Misty's only 1 year and so is Tomodachi Joto. That one's my favorite.

Alani: Look, they're both pretty light—Misty weighs 9 and Joto weighs 6.5.

So how do I keep track of how much they weigh?

Lindsay: Just put the weight right next to the age, like you were doing before.

Continued on next page

Trevor: I'd do it a different way.

First, let's make sure we understand what Alani and Noah and Lindsay are doing. What would you do next?

Noah: We'd just keep listing the cats in order of age. If their weights are right next to their ages, we can see how the weights change when they get older.

Okay, that's a good method. Trevor, what were you going to say?

Trevor: I think we should divide the cats in half—into old ones and young ones, then see if the weights are different.

[The teacher writes Old Cats and Young Cats on the board.]

Which cats go in each category?

Trevor: Well, let's just say anything more than 6 years goes in old, because 6 is a middle-aged cat.

Okay. Tell me what to do next.

Mei-Ling: Just write in each cat and its weight in the right place. So write "Pepper" and "12 pounds" in Young Cats.

Trevor: And "Strawberry, 14.5 pounds" in Old Cats.

Okay, we have a couple of ideas to get us started. You can use one of these representations or find a new one to see how your information goes together.

Using the Computer to Investigate Cats

Materials

- Database software, such as Tabletop Sr.
- Computer (at least 1)
- Cat data collected by students (Student Sheet 7)

What Happens

Students add their own cat data to a computer database and use it to explore questions about the relationships between variables. Their work focuses on:

- using a computer to examine data
- learning the operation of a computer database tool
- framing and exploring questions about relationships between variables

Activity

Exploring with Database Software

Before You Begin If you have data for only 24 cats (the Cat Cards), it isn't so hard to explore the data on paper, as students have been doing. But as more cats are added to the sample, or as more questions about the cats are addressed, keeping track of the information becomes more difficult. With database software, it is easy to find all the green-eyed cats, or all the cats that weigh more than ten pounds, in a very large sample.

An ideal database for this Excursion is Tabletop Sr., available from Dale Seymour Publications in either Windows or Macintosh versions. The hardware requirements are as follows:

Windows: 386SX or higher; Windows 3.1; 4 MB RAM; VGA

Macintosh: System 6.0.8 or higher; 2 MB RAM; 256 color monitor

The Tabletop Sr. database "Cats" corresponds to the cat database on the cards. If you use different database software, you or your students will need to key in the data from the Cat Cards.

Tabletop Sr. allows the user to ask relationship questions of the database (for example, "Are older cats heavier?"). Icons representing the data move around the screen to display the answer. With Tabletop Sr., you can select from many representations, including Venn diagrams, scatter plots, and graphs. Other database software suitable for elementary school may offer fewer options.

For any database software, see the accompanying user's guide and allow some time for students to become familiar with its features.

Exploring the Cats Database Depending on the number of computers you have available, it may take a few days for all students to explore the cat database. Here are some things students might do both to learn the database program and to study the cats:

■ Begin by encouraging students to find data that is of interest to them. They might have a database scavenger hunt. On slips of paper, list things to find, such as all the gray cats, all the female cats, and all the cats that are over age 8. If your software allows you to look for the intersection of two variables (as Tabletop Sr. does), students can also find groups of all the cats that are both gray and over 8 years old, or all female cats that weigh more than 8 pounds.

■ Students can enter their own cat data in a new file or add it to the original sample of cats. If they use a separate file, they can compare their own cats to the original sample of cats along any of the characteristics. For example: How do our cats compare to the original set of cats in terms of their ages? their weights? their lengths? If students add their data to the original database, they can see if the conclusions they drew from the original data still hold. For example: Is it still true that more than half the cats are older than 7 years?

■ Using Tabletop Sr., students can try a game made up by a fifth grade class: Choose one particular cat. Find some number of intersecting circles so that this particular cat is alone in the "middle," or in the intersection of all of the circles. If students have added data about their own cats, it is particularly interesting for them to try to get their own cat in the middle.

■ With Tabletop Sr., students can experiment with different representations to address the association questions they raised in Session 2. For example, they may want to use a scatter plot to find out "Are older cats heavier?" They choose values for the axes on the graph, then watch as each cat icon moves to the appropriate coordinate for its age and weight.

Sampling Ourselves

What Happens

Session 1: Review: Fractions, Decimals, Percents Students collect data on questions that have two possible answers. Using these data, they review several methods to change data fractions to familiar fractions, decimals, and percents. They use numerical reasoning, Data Strips, and calculators to do these conversions.

Sessions 2 and 3: Sampling the Classroom Students sample themselves, using their class as a population and a group of four as a sample. They carry out a survey of six questions, then work in small groups to see how their group does or does not represent the entire classroom in its response to a particular question.

Session 4: The Classroom as a Sample Students use correspondences among familiar fractions, decimals, and percentages to see how their class is like or unlike a larger sample of 8- to 12-year-olds who participated in a national survey. They discuss why data about their class are similar to or different from data gathered from the national group.

Mathematical Emphasis

- Learning what a sample is and some of the factors that make a sample reasonable
- Comparing the data from a sample to the data in a larger population using fractions, decimals, and percents
- Using data characteristics to compare a sample with a larger population
- Learning why a larger sample tends to reflect a population better than a smaller one

18 out of 29 in our class is close to $\frac{2}{3}$ – but a little less.

What to Plan Ahead of Time

Materials

- Adding machine tape: 2-foot strip for demonstration (Session 1)
- Scissors: 1 per student (Sessions 1–4)
- Tape or glue as needed for Data Strips (Sessions 1–3)
- Calculators: 2 per small group (Sessions 1–4)
- 3-by-5 cards or slips of paper: 6 per student (Sessions 2–3)

Other Preparation

- Duplicate student sheets and teaching resources (located at the end of this unit) as follows:

For Session 1

Data Strips (p.150): 1–2 sheets per student

Student Sheet 8, Finding Familiar Fractions (p. 146): 1 per student, homework

For Sessions 2 and 3

Data Strips (p. 150) 1 sheet per student, plus some extras

Student Sheet 9, Small-Group Sampling (p. 147): 1 per student

For Session 4

Student Sheet 10, Meals and Chores Survey (p. 148): 1 per student

Student Sheet 11, Survey Results (p. 149): 1 per student

Session 1

Review: Fractions, Decimals, Percents

Materials

- Adding machine tape (2-foot strip for demonstration)
- Data Strips (1–2 sheets per student)
- Scissors (1 per student)
- Tape or glue (as needed)
- Calculators (2 per small group)
- Student Sheet 8 (1 per student, homework)

What Happens

Students collect data on questions that have two possible answers. Using these data, they review several methods to change data fractions to familiar fractions, decimals, and percents. They use numerical reasoning, Data Strips, and calculators to do these conversions. Student work focuses on:

- estimating the size of fractions using numerical reasoning
- converting unfamiliar fractions to familiar fractions
- finding decimal and percent equivalents to fractions

Note on This Review Session

This investigation starts with a review of three methods of changing a data fraction like "14 out of 27 students" to a familiar fraction, decimal, or percent. These are the methods:

- Using numerical reasoning about the fraction (for example, $15/31$ is close to $\frac{1}{2}$ because 2×15 is very close to 31)
- Creating a Data Strip for a data fraction and comparing it to a fraction strip folded to show familiar fractions
- Using a calculator to divide the number in a category (for example, 13 boys in a class) by the total number in the sample (for example, 31 students total in the class) to get a decimal or percent

This session is a review of the content of the grade 5 Fractions, Percents, and Decimals unit, *Name That Portion*. If your class has not done that unit, plan to spend more time on this session before you proceed.

The first two activities in Session 1 help students use Data Strips and calculators to work with data fractions. There is not a separate activity on numerical reasoning about data fractions, but encourage students to use such reasoning whenever they can, both to find familiar fractions and to check the answers they get with Data Strips and calculators. It is important not to let the strips and calculators distance students from their sense of the numbers. At times during the next two activities, you might ask part of the class to figure out a familiar fraction with numerical reasoning, while others work with strips or calculators.

Using Data Strips

To gather some data to work with, take a poll on students' TV preferences. Ask students for the name of a current TV show, then do a quick survey to find out how many students in the class watch the show and how many do not. On the board, record the two numbers and the total, leaving room for additional information after each. For example:

Watch	Do not watch	Total
[TV show] 11	[TV show] 14	in class 25

Ask students what fraction of the class is in each category, and write it next to the data:

Watch	Do not watch	Total
[TV show] 11 $\frac{11}{25}$	[TV show] 14 $\frac{14}{25}$	in class 25

When people report on data like this, they don't usually say *[substitute your own data here],* **"11 out of 25 people watch this TV show." It's hard for most people to understand 11 out of 25. People can understand better if data are reported as a familiar fraction, such as about ½, or a percent, such as 44 percent, or a decimal, such as 0.44.**

In the rest of this unit, you will be collecting, reporting, and comparing a lot more data. We'll review a few ways to make complicated fractions like 11/25 easier to understand. For the first way, you will fold strips to show fractions.

Demonstrate the process of making a fraction strip. If students have done the *Investigations* unit *Name That Portion,* they can add to your demonstration by describing their previous experience. Fold a strip of adding machine tape into halves, thirds, fourths, and sixths. When you are done, your strip should show ⅙, ¼, ⅓, ½, ⅔, ¾, and ⅚. Label each fold with the corresponding fraction.

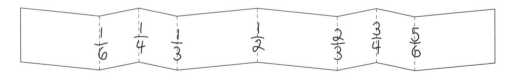

Now students make their own strips. Give each student the sheet of four Data Strips.

These are called Data Strips. Each block on the strips stands for one person in the class, so first you need to make two strips that are the right length for our class. Don't include anyone who is absent today; we need strips that have one section for each person we just counted in our poll.

Students cut out two strips and adjust them so that each strip has one section for each student in the class. There are 30 sections on each strip, so if your class has fewer than 30 students, the extra sections should be cut off. If your class is larger than 30, the students will have to tape or glue additional blocks (part of another strip) to each strip they are preparing. When they are finished, students should have two identical strips, each strip with as many blocks as there were students answering the TV poll.

One of these strips will be your fraction strip. It will show the familiar fractions, which you can compare to your data fractions. Take one strip and fold it into halves, thirds, fourths, and sixths. Then, *on the blank side of the strip,* label each fold with its fraction.

Allow students a few minutes to fold and mark their fraction strips. You may need to help them with some of the labeling, especially for fractions that do not have 1 in the numerator. The back of a finished strip should look like your demonstration fraction strip (p. 61). Some students may also want to mark eighths, since they are easy to fold from fourths. This would add 1/8, 3/8, 5/8, and 7/8 to the strip.

Now use your other strip to show our data on students who watch the TV show: 11/25. Put an X in a section of the strip for each student who does watch the TV show.

You may want to mark a sample strip to demonstrate.

Which sections of your strip show the people who do not watch this show? How could you use this strip with your fraction strip to figure out what familiar fraction 11/25 is close to?

Students work briefly in small groups to find a familiar fraction close to 11/25 by comparing their fraction strip and their Data Strip. Be sure they align the filled-in part of the Data Strip with the lower end of the fraction strip. Record students' answers next to the original data. More than one answer will be correct if a data fraction lies about the same distance from two familiar fractions.

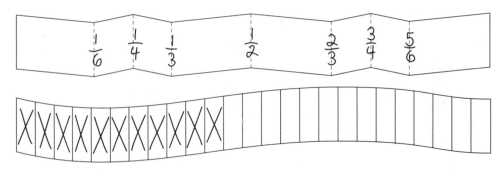

Fraction of people who watch

Now use your fraction strip and the same Data Strip to find a familiar fraction for those who do not watch the TV show.

Be sure that this time students align the part of the Data Strip *not* filled in with the lower end of the fraction strip, as shown at the bottom of this page. Record these results as well.

Note that each time students work with a different denominator, they will need to cut and fold a new fraction strip. If they are working with 13/22, for example, they will need a fraction strip that is 22 sections long to compare with their Data Strips. While making a new fraction strip is an extra step, students will not have to make a new one for every set of data; if they are dealing mostly with data from their own class, the denominator will not often change.

When considering your students' answers, remember that finding a familiar fraction to express a data fraction is an estimation process, so often there is not just one correct answer. For example, the fraction 15/21 is between 2/3 and 3/4 and is approximately the same distance from each. Some students may reason that 21 is close to 20, so 15/21 is close to 15/20 or 3/4. Others may reason that 15 is close to 14, so 15/21 is close to 14/21 or 2/3.

The strips will not help students decide which is closer, and either answer is a reasonable approximation. Encourage this kind of reasoning, but realize that in many cases it will be beyond students' knowledge of fractions to judge which estimate is closer to the actual fraction.

It is also possible that even when everyone in the class answers a question yes or no, the familiar fraction of the class that answers yes and the familiar fraction of the class that answers no will not add up to 1. For example, the familiar fraction for 17/28 could be 1/2, and that for 11/28 could be 1/3. But 1/3 and 1/2 do not add up to 1. This happens because 17/28 is really a little more than 1/2, and 11/28 is really a little more than 1/3. Students may think they have made a mistake when this happens. Discuss why this is *not* an indication of an error.

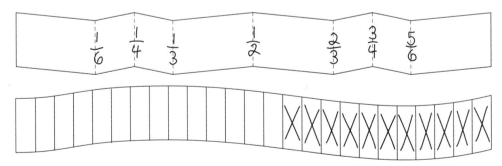

Fraction of people who do not watch

Using a Calculator for Decimals

Collect data on a new question so you have a new data set to work with. You might ask, "Do you prefer your pizza with or without olives?" (or another question that has two possible answers). Write the data on the board in numbers and fractions, as before.

❖ **Tip for the Linguistically Diverse Classroom** If you ask the pizza question, provide pictures (pizza ads are a good source) or a can that illustrates the topping under consideration.

This time we'll change the data fractions into decimals and percents.

If students have worked through *Name That Portion,* remind them of the table they filled out that showed how to make a fraction into a decimal. Otherwise, go directly to the calculator.

How would you use the calculator to get a decimal for the number of people who prefer their pizza with olives?

If we wanted to use a percent instead of a decimal, what would that number be?

As necessary, remind students that the first two numbers after the decimal point are the approximate percent, for example, 0.245 is about 24%; and that a percent with only one number is the equivalent of a decimal with a zero after the decimal point, for example, 8% = 0.08.

Using their calculators, students next find the decimal and percent for the fraction of students in their class who do *not* like olives on their pizza.

Working with Class Poll Data

Collect another data set, using a different question, such as "Are you wearing shoes that tie today?" As before, record the data for the yes and no responses. In small groups, students find a simpler representation (familiar fraction, decimal, or percent) for these fractions, using their choice of Data Strips, calculators, or numerical reasoning.

Ask a few to share their results and their strategies. If there is some method that no one has used, you may want to demonstrate it, or wait until the next data collection task to see if anyone uses it then.

Help students focus on the process of examining data by asking questions such as the following:

Did any of you come up with different familiar fractions for the same data fraction? Why might this happen?

The two data fractions $10/27$ and $17/27$ add to 1. Are there examples of two familiar fractions for these data fractions that do not add to 1? Why might this happen? Are there examples of two percents for these fractions that don't add up to 100 percent? Why might that happen?

Continue by choosing another topic for a poll, collecting data, finding familiar fractions, decimals, or percents, and examining them. Until your students are comfortable with this process, continue to use polls that have only two possible answers. Then try taking polls with more than two answers. Following are some questions with more than two answers; students might also suggest questions of their own.

Are you the oldest (or only), youngest, or in the middle in your family?

How do you get to school—walk, bus, bike, car, or another way?

What is your favorite carbonated drink?

At the end of class, collect data for three more questions that students will work on for homework. Hand out Student Sheet 8, Finding Familiar Fractions, and give students time to record the data on this sheet before taking it home.

Session 1 Follow-Up

Homework

Students complete Student Sheet 8, Finding Familiar Fractions, for the data they recorded in class. Send home extra Data Strips sheets for students who want to use them.

Students who have calculators at home can extend the activity by also finding decimals and percents for each data fraction.

Sampling the Classroom

Materials

- Data Strips (1 sheet per student, plus extras)
- Scissors (1 per student)
- Tape or glue (as needed)
- Calculators (2 per small group)
- 3-by-5 cards or slips of paper (6 per student)
- Student Sheet 9 (1 per student)

What Happens

Students sample themselves, using their class as a population and a group of four as a sample. They carry out a survey of six questions, then work in small groups to see how their group does or does not represent the entire classroom in its response to a particular question. Student work focuses on:

- learning what a sample is
- figuring out ways to compare the data from a sample to the data from a larger population
- considering how the size of a sample affects how well it represents a population

 Ten-Minute Math: Volume and Surface Area During the remainder of this unit, use the Volume and Surface Area activity for Ten-Minute Math. This activity is related to experiences in the grade 5 unit *Containers and Cubes* (3-D Geometry: Volume). If your students have not worked on that unit, you may want to continue with The Digits Game for Ten-Minute Math for the remainder of this unit.

Before doing the Volume and Surface Area activity for the first time, cut apart the five groups of images on the transparency (p. 154) and store each group in an envelope with its letter on it.

Introduce this activity by showing students a cube and asking them to imagine a square stamp that is the same size as one of the cube's faces. Ask them how many square stamps it would take to completely cover the cube.

Then show figure A-1 on the overhead projector.

How many stamps would it take to cover the outside of this cube solid? How many cubes make up the solid?

Have cubes available for students to figure or check with. Share strategies. You can keep a record for each cube figure and look for patterns across related configurations.

Reviewing the Sample of Cats

Yesterday we asked the question, "Do you prefer your pizza with or without olives?" We found what was true for our class. Suppose we wanted to find out how all the people in a city, or in the whole country, felt about olives on pizza?

It would take a very long time and cost a great deal of money to ask everyone. What might we do to get an idea of what we would find out if we *could* ask everyone?

Accept ideas from students about approaches to this problem. Elaborate on any suggestions that have to do with sampling. You might ask students if they have heard reports on the results of surveys. Ask what the surveys were about, and who was questioned.

Quality control is another real-life example of sampling that you might want to describe to students:

Sometimes, in businesses, people use samples in order to know if they are reaching a goal. For example, suppose you work in a factory that makes wind-up toys. Your goal might be for 99 percent of the toys to work perfectly. You wouldn't have enough money to test every single toy, but you might choose a sample of toys and check to see what percentage of the sample toys were working perfectly. This is sometimes called quality control; it's another way to use samples.

We're going to spend a few days figuring out how to use a sample—a part of a group—to find out about a larger population.

Remind students of the sample they worked with in Investigation 2.

We've recently looked at a sample of cats—just a small number of cats, compared to all the cats in the world. In what ways do you think the sample we studied was like the whole population of cats? In what ways was our sample probably different from the whole population of cats?

Spend a few minutes reconsidering the sample of cats. Students might have noticed some kinds of cats missing from the sample. For example, there were no Siamese cats. There were no all-white cats. And there were no kittens, so the sample was biased in the direction of older cats.

The cats on the Cat Cards probably were not representative of the whole population of cats. Our sample doesn't have all the kinds of cats in the population, and it may have more of some kinds of cats than you'd expect in the whole population. As a way to get useful information about all cats, it really wasn't a very good sample.

If your students have collected and analyzed their own cat data, ask them if they think their extended sample is representative of the whole population of cats. Does adding their cats to the original set of cat data make the set more representative?

In preparation for the next activities, encourage students to think about ways of sampling a population of cats. Point out that the populations they can easily sample may not be as extensive as they first think. That is, they may be sampling only domestic cats in our town, or domestic cats in our community, rather than cats as a whole population.

Activity

Sampling the Class

It's important to be careful when taking a sample so that we'll really be finding out about the whole group. One question is, How big does the sample have to be? We'll explore this by doing some sampling with our class.

Let's suppose that we have a question, but we can't ask everyone in the class. We can only get information from a sample of the class. We'll try asking a sample of four students, and then figure out from that sample what we think is true about the rest of the class. Later, we'll talk about whether or not four is a good size sample for our class.

The first question we'll try to answer is this: What fraction of our class is right-handed? We'll define right-handed as *writing* with your right hand, even if you eat or throw with your left.

Randomly select four students to be the sample (for example, by choosing the first four students on your class list), and have them sit or stand together. Ask them to raise their right hands if they are right-handed, and record this information on the board; for example, "3 out of 4 people in the sample are right-handed," or "All 4 students in the sample are right-handed."

Based on this sample, what fraction of the whole class would you predict is right-handed? About how many students is that?

Allow a little time for students to discuss their inferences and methods. Encourage strategies that use familiar fractions. For example:

> I knew that ¾ of the sample was right-handed, so I had to figure out what ¾ of the whole class was. If you divide 25 into quarters, it's about 6. So ¾ of the whole class would be about 18 kids who are right-handed.

Now collect the data for the whole class. How many people in the class are actually right-handed? Write the data from the sample and the data from the population on the board.

Is our sample like the population?

Students may use any of the tools they reviewed in Session 1 to make the comparison between the sample and the class population. After students make their comparisons, they share their methods. If any of the following methods are not suggested, describe them yourself. It is important for students to be able to work with fractions, decimals, and percents in order to understand data that are reported in any of these ways.

- Use a Data Strip and a fraction strip to find a familiar fraction. Make a fraction strip for the number of people in the class. Make a Data Strip representing the right-handed people in the class. Find a familiar fraction for the class and compare it with the sample fraction.
- Find decimals for both fractions, using the calculator, and compare them.
- Find percents for both fractions, using the calculator and changing the decimals to percents, and compare them.
- Use numerical reasoning to approximate a familiar fraction, decimal, or percent from the class fraction and compare it with the sample fraction.

As students address the question about the sample and population, encourage them to make comparative statements. For example:

> About ¾ of the population was right-handed, but only ½ of the sample was right-handed.

Write students' statements on the board. Students will probably wonder how close the fractions have to be in order to consider them similar. While there is no absolute answer to this question, you might use as a rule of thumb that the class fraction should be closer to the sample fraction than to any other fraction with fourths. Briefly consider whether this sample was informative for finding out about the class.

Small-Group Sampling

Distribute six 3-by-5 cards and a new copy of the Data Strips sheet to each student. Form groups of four students each. If your class size is not a multiple of four, put five in some groups. If your class is smaller than 24 students, use fewer survey questions, so that no groups have fewer than four students.

We're going to do a survey. We'll think of each group as a sample of our class population. Each group will see how well they represent the population on one of the survey questions. Some of the questions are serious, and some are silly.

First, we'll all respond to the survey. Number your cards from 1 to 6. Answer each survey question on the corresponding card. For example, the answer to question 1 goes on card 1, and so on.

Read the questions to the class, making sure to give the question number each time. Ask only as many questions as there are groups of students. Encourage students to keep their data private.

❖ **Tip for the Linguistically Diverse Classroom** Have visual aids available to make each survey question comprehensible, and print the answer choices on cards. Advertisements are a good source of pictures, or have actual items if possible. For example, for question 3, you might demonstrate or draw, step-by-step, a sandwich made both ways. Match these examples with cards that read *separate* and *same.*

Question 1: Do you wear a seat belt when you are riding in the car? (Write *yes* or *no.*)

Question 2: Do you squeeze toothpaste from the bottom or the top of the tube? (Write *top* or *bottom.*)

Question 3: When you make peanut butter and jelly sandwiches, do you put the peanut butter and jelly on separate bread slices, or on the same slice? (Write *separate* or *same.*)

Question 4: When you put on your shoes and socks, do you put on both socks, then both shoes, or do you put on one sock, then one shoe on each side? (Write *both* or *1 sock, 1 shoe.*)

Question 5: Do you put your ketchup on top of french fries or on the side? (Write *top* or *side.*)

Question 6: When you eat cookies with a filling in them, do you take them apart or not? (Write *take apart* or *keep together.*)

Students will probably want to know what to do if they can't answer a question, for example, if they don't eat french fries or don't eat ketchup on them. Tell them that when this happens in a real survey, they would answer "not applicable," abbreviated N/A. This means that they aren't part of the population or the sample for this particular question.

Another data collection issue is that students sometimes feel they could choose either answer to a question. For example, some students have said that they squeeze toothpaste from the top when it's new and from the bottom when it's old. In these situations, students should answer whatever they do more of the time. It is important that everyone in the class use the same rules for answering the questions so that the answers are comparable.

When students have written their answers, assign a question to each group. If there are more than six groups, assign the same question to two groups. Tell students that each group is going to be a sample of the classroom population for their question. Each group records its own question and how each member of their group responded to it on Student Sheet 9, Small-Group Sampling. The three rows for the responses in each chart include a place to record N/A answers.

Each group then uses their information to make a prediction about the class population and records it on the student sheet.

Now, students compile the class data. Gather all the cards in six separate piles, one pile per question. Label the piles Question 1, Question 2, and so on, to avoid any confusion. When all the data are in, each small group collects the class data for its own question. If two groups have the same question, they can work together to compile the data. They count the responses and record them on the student sheet. Working in their groups, students use Data Strips, calculators, or numerical reasoning to compare the fraction in their group sample to the fraction in the class population.

As students are working, make a simple chart on the board for groups to share their findings as they finish. With data added, the first two rows of the chart might look something like this:

	Sample	**Population**
Question 1	3/4 wear seat belts	24/26 wear seat belts
Question 2	1/2 squeeze bottom of tube	15/24 squeeze bottom of tube

It is important that students express both sample and population data in terms of the same option (e.g., bottom of tube) so they can compare them. Be sure each group has done this before recording their data on the chart.

When everyone has finished, each group explains to the rest of the class whether their group sample was very much like the population or not. At this point, some students may state their sample findings in terms of "right" and "wrong." Explain that the similarities and differences they are seeing between sample and population are merely the result of the sampling process, not of good guesses or correctness on anyone's part. Wrap up this discussion by asking students more general questions about the samples and populations.

Which samples were like the class population? Which were different from the class population? Were all the samples representative of the whole class? Why or why not?

As a check on individual students' understanding, have each student write a response to the last question on Student Sheet 9.

The main purpose of this discussion is to develop students' ideas about the importance of the *size of the sample*. It is likely that some of the samples were very unlike the class population. All or none of the students in a sample may have responded a certain way, but it would be rare for this to happen in the class as a whole. Samples of four are seldom large enough to give us good information about a population. But there are some samples that are very good at representing a larger group. It is important that students see this so that they will understand that sampling can be a useful method of gathering data. See the **Teacher Notes,** Whys and Hows of Sampling (p. 74) and Samples Vary (p. 75) for more information on sample size.

Would we be able to predict what's true in the population more accurately if we used a bigger sample? Say, if we sampled about half the class rather than four people?

After students discuss this possibility, try it out with one of the questions. Ask for a volunteer to randomly choose about half of the cards with students' responses to one of the questions. Quickly record the results.

Was this sample more like the class population than the original sample was? Why do you think that happened?

Try drawing a bigger sample for other questions as well and see what happens. Most of the time (but not always) the bigger samples will be closer to the whole class than the smaller samples were.

Finally, since the best sample would be the whole class, review why samples are useful to begin with and what the trade-offs are between taking larger and smaller samples.

Why do people take samples instead of asking everyone? How might they decide how large a sample they need?

Sessions 2 and 3 Follow-Up

 **Homework**

Students find an article in a newspaper or magazine that reports on a survey in which only a sample of a population was questioned. It must be an article they can understand and explain. They read the article and write a paragraph telling (1) what the survey tried to find out and (2) what the sample was, if it is reported in the article. Students may find that some news reports do not include details about the sample. Their experiences with samples so far should lead them to question why this information was left out and how the selection of the sample might have affected the results of the survey.

❖ **Tip for the Linguistically Diverse Classroom** Offer these alternatives for completing the assignment:

■ Pair any student who has limited English proficiency with an English-proficient student who lives nearby, who can read the article aloud and help record the partner's answers.

■ The article and the students' responses may be written in the students' primary language. They may also have help reading the article.

This introduction to sampling is closely related to the work on probability in the grade 5 unit, *Between Never and Always*. The role of sampling in these two contexts is similar, although the population being investigated differs. In the probability setting, there was no clearly defined population; just a sense of, for example, how often a spinner would spin green *in the long run*. In this data setting, there is a defined population, for example, all the cats in the world or all the children who use playgrounds in the United States, or all people between the ages of 13 and 19 in the United States.

In both settings, we are trying to get an idea of how something works in a large number of cases by looking at a smaller number of cases. We do this because it would be impossible or too expensive or too time-consuming to look at every single case. So sampling is a way of cutting down our work.

Here's an analogy that comes quite close. Suppose we wanted to buy the right amount of fabric to cover a couch. The most accurate (and the hardest) way would be to measure every little piece of the couch to be covered. Or we could measure just a few parts of the couch and estimate. The more parts we measure, the more accurate we'll be, and the more time and effort it will take. But if we take too few measurements, we may get the wrong amount of fabric.

To continue the analogy, two aspects of how we take measurements might affect the quality of the information we get. The first is *how much* of the couch we measure—in general, the more we measure, the closer we will come. The second is *which parts* of the couch we measure. If we measure just the shorter parts and use them to estimate the total measurement, we won't do as well as if we use a representative sample of measurements—some long, some short—from different parts of the couch.

These two characteristics of sampling are formally known as *sample size* and *sample representativeness*. The significance of sample size is somewhat obvious: The bigger the sample, the better the information about the population is likely to be. Very small samples may provide very little information; some of the four-person samples your class took in Investigation 3 probably demonstrate this difficulty.

On the other hand, statisticians have found that there are diminishing returns as sample sizes get bigger. You can get a sense of this by thinking about how much information you can get by taking a sample of everyone in your school except 10 people, or everyone except 20 people. These two samples would be almost identical in the information they provide, even though one would take longer to collect. The size of national samples are chosen to be reasonably easy to collect, but also to give good enough information that the next survey won't come up with totally different results.

Sample *representativeness* refers to the idea that the sample must be similar to the population in order to provide accurate information about it. For example, if you are interested in the general health of students in your school, you might survey everyone who was in school one day. But you would be missing those students who were out sick, and thus you would miss important information about student health. This would be a somewhat unrepresentative sample.

In practice, representative samples are generated primarily by (1) assuring that there are no obvious dissimilarities between population and sample and (2) choosing the sample in a way that will not create any dissimilarities. In the activity on sampling for left/right-handedness, it might seem that any set of four students in the class will be similar to the whole class. But

Continued on next page

suppose you had desks at the left end of the rows especially for left-handed students, and you created samples by grouping all the students at the left end of the rows, all the students one in from the left, and so on. You would end up with one sample that was all left-handed and several that were all right-handed, because the sampling procedure itself created a dissimilarity between the sample and the population. These samples would not provide good information about the whole class.

We can never be totally sure that a sample is representative. The best we can do is to avoid any obvious problems in choosing the sample. Always ask first if there are any striking differences between sample and population. Random methods of choosing a sample usually work well. By drawing names out of a hat or grouping by initial of first or last name, we usually avoid introducing any bias into our sample that would prevent our using it to find out about a population.

Samples Vary

A most important point to remember throughout this unit is that samples vary, *no matter how carefully we choose them.* Here is an example of how this fact might confuse students:

One class took a sample of a population of 500 red, blue, and yellow marbles. In the first sample of 100, they had 41 blue, 19 yellow, and 40 red marbles. Because there were about twice as many blues and reds as yellows, the students decided that there would be 200 blues, 100 yellows, and 200 reds in the population.

Confident, the students replaced the marbles from the sample, remixed them, and took another sample of 100. This time there were 32 blue, 21 yellow, and 47 red. Based on multiplying each number by 5 (because there were five times more marbles in the population than in the sample), the students now thought there would be 160 blues, 105 yellows, and 235 reds.

This completely threw the class, because the blues and reds no longer seemed close to being equal; in fact, there were about 50 percent more reds than blues. One student suggested that they

should average out the results between the two samples. Another suggested that they take three more samples and then decide. A third wanted to count the population to decide which sample was right.

The teacher, who knew that the first sample yielded the "correct" information, was unsure how to proceed. She remembered that samples could vary and could yield a variety of results, so she waited for the students to talk about their reactions and then encouraged them to take three more samples. These were very similar to the students' first sample, and on the basis of that, the students were content to keep to their original prediction.

The students welcomed the information that there was variability in samples, and they understood that sometimes the results are "out there," as one student termed it.

Session 4

The Classroom as a Sample

Materials

- Student Sheet 10 (1 per student)
- Student Sheet 11 (1 per student)
- Scissors (1 per student)
- Calculators (2 per small group)

What Happens

Students use correspondences among familiar fractions, decimals, and percentages to see how their class is like or unlike a larger sample of 8- to 12-year-olds who participated in a national survey. They discuss why data about their class are similar to or different from data gathered from the national group. Students' work focuses on:

- changing class data fractions into percentages
- using characteristics of two samples (one small, one large) to understand similarities and differences between them
- exploring whether a sample is representative of a larger sample or population

Activity

Are We a Good Sample?

In the last sessions, we looked at how well a small sample of students in our class represented the whole class. Now we're going to look at our whole class as a sample of a larger sample of young people. We're going to compare our responses to survey questions with the responses of 1004 young people, 8 to 12 years old, who participated in a national survey.

Note: The activities in this session draw on data from a national survey, "America's Children Talk About Family Time, Values, and Chores," sponsored in 1994 by the Massachusetts Mutual Life Insurance Co. Because these data are from a survey of 1004 young people in the United States, they are *not* data on the population itself, but on a large sample. In fact, we almost never have data on everyone in a country, except in a census report.

Even though these data are from a sample of the whole population, the sample is large enough that we assume it gives us good information about the population. However, it is important to be clear with students about the difference between the large sample and the population, and to refer to the original group of survey subjects as a "large sample."

Distribute Student Sheet 10, Meals and Chores Survey.

❖ **Tip for the Linguistically Diverse Classroom** Read each question aloud, allowing time for students to respond to each question as it is read.

Students complete the survey and cut their responses into strips. Gather the strips into seven piles, one per question.

When we count these responses, we'll have data for our class like this: "14 students out of 25 do not help set the table." The data from the larger survey are reported in percents. What could we do to compare our class to the large sample?

One obvious approach is to change the class data to percents. Another possibility is to convert both data reports to fractions or decimals. Some students may suggest using the percents from the national survey to find the number of students in the class that would be the equivalent of each percent. Any of these techniques will work.

The next activity assumes that students will be changing the class data fractions to percents. If some of the other techniques are interesting to your students, they can use them in parallel to that method.

Comparing the Survey Results

Write on the board:

> What percent of students in our class gave each response to the question?

Divide students into seven groups, one per question. Give each group the collected strips of data from one question.

Each group will work with one question. You can use any of the methods we've been working with to figure out the percentage: finding a familiar fraction and using the equivalent percent; using the calculator to get a decimal, then changing it into a percent; or doing mental math to estimate a percent. Just be sure you can say how you got your answer.

Each group figures out the approximate percent of the class that gave each of the three responses to their question. As students finish their work, they record their findings on the board. They write their question number, the possible responses, and the percent of students in the class who responded each way.

What is most interesting about these results? Are there answers that surprised you? Do you think our class is a typical class?

Now tell students details about the national sample:

The 1004 young people who responded to the survey included an equal number of kids 8, 9, 10, 11, and 12 years old. Equal numbers lived in the northeast, south, midwest, and west. Half were girls and half were boys. About 80 percent lived with both parents, and the rest lived with one parent, usually their mother. In half of the families, both parents worked.

Question 3

breakfast 20%
lunch   44%
dinner   36%

Question 6
yes 12%
~~sometimes~~

Distribute Student Sheet 11, Survey Results. As students copy their class results from the board onto this sheet, discuss how your class compares with the larger group who answered the questions in the national survey.

Are you surprised at any of the survey responses? In what ways are we most different from the survey students? What do you think might be a reason for these differences? Is our sample more like or more unlike the national sample?

Encourage students to talk about factors that might have made their responses different from the national sample. For example:

> Almost all of us have two parents who work, and only half of them did. Maybe that's why we don't eat dinner together as much.

> We're older than a lot of those kids. We have more responsibilities. That's why we probably do more cooking.

> A lot more of us live with only our moms. We have to help out more, so we do more cooking.

Students may make comments about the differences between their class and the larger sample that are unrelated to the data, such as "We all go to the same school and they don't." Remind them to first compare the class and national percentages, then comment on how the characteristics of the groups might or might not account for the similarities and differences between the data. See the **Dialogue Box,** Why Is This Sample Different from the Population? (p. 81).

Note: Students may wonder why some of the responses to the national survey questions don't add up to exactly 100 percent, but to 99 or 101 percent. This is an interesting mathematical question. If students raise it, ask them to think about how these percents were calculated. Did the students have any percents or decimals that didn't add up to 100 percent or 1.0 in their results? Why might this happen? Can they figure out a set of decimals that add to 1.0 but that, when they are rounded, add to 0.99 or 1.01?

For homework, students will write about the comparison between their class data and the national survey data, so be sure they have copied their data from the board onto Student Sheet 11 before they leave class.

Session 4 Follow-Up

 **Homework**

Students write about how their own results on the Meals and Chores Survey compared with the national results, referring to the data on their completed copy of Student Sheet 11.

You may want students to write about the question they worked with in class, or let them decide which questions to tackle. Listed below are points students should cover. You may wish to write these on the board for students to copy. If students are accustomed to open-ended assignments, simply ask them to write their own comparison of the data.

- How were our class results the same as or different from the national survey results? Which of these similarities and differences surprised you?
- What could be the reasons for the differences?
- According to the data, is our class a representative sample of the national survey group? Why or why not?

❖ **Tip for the Linguistically Diverse Classroom**   Students who are not yet writing comfortably in English may respond in their primary language.

Why Is This Sample Different from the Population?

During the activity Comparing the Survey Results (p. 77), these students are discussing how their classroom results differ from those of the national survey.

Marcus: The national results were completely different. In our class, one-tenth helped cook, and in this one, one-fourth did. That surprised me.

Jasmine: Our parents probably don't trust us as much to use the stove, so we don't get to cook as much.

Duc: I was surprised by the favorite meal. We thought lunch, with like 40 percent, but they thought dinner, with 43 percent.

What percentage of them liked lunch best?

Duc: It was 33 percent. It's not too much different, but it's still different.

Corey: Probably it's because a lot of kids don't like school lunches. We don't have school lunches, so we get to pick what we like for lunch.

Jasmine: That's probably why lunch is our favorite!

That reminds me about the question on whether you get to choose what you eat for dinner. What did you find there?

Natalie: We didn't get the same as they did. I don't think we did it right.

Well, it's not about finding the same thing. We might be different than they are for some very good reasons. How were we different on this question?

Natalie: Forty percent of them didn't get to choose what they wanted for dinner. Only ten percent of us didn't get to choose.

Mei-Ling: Two-thirds of us got to choose at least sometimes, but only one-third of them got to choose sometimes.

Jasmine: Yeah—we get to choose more often, but they get to cook more often.

Our group is a little older than the national group. Do you think that would make a difference on any of these questions?

Mei-Ling: It's pretty much the same range. They just have some younger kids. Maybe that's why they don't get to choose what they want to eat as much. We're older so we have more choices.

A Sample of Ads

What Happens

Session 1: Fractions of Newspaper Pages
Students practice finding the fraction of a newspaper page that is made up of ads, using a variety of strategies. They describe and compare their strategies.

Session 2: Collecting Data from Ten Days
Students define a sampling strategy so they can take data from their newspaper in a single class session. Each small group figures out proportions of ads on a sample of pages in a single day's newspaper and records the data.

Session 3: Combining Ad Data Across Pages
Each small group combines their data to calculate a single fraction for the entire paper. Then the groups compare their results and discuss their sampling strategies. Students compare their data to the newspaper's stated goal for ads.

Mathematical Emphasis

■ Choosing a strategy to find a representative sample

■ Figuring what fraction of a newspaper page is covered by ads

■ Comparing data from a sample with a target fraction

■ Combining fractions of pages into a single fraction of a newspaper

■ Analyzing a single data set in more than one way

What to Plan Ahead of Time

Materials

- Rulers or tape measures: 1 per small group (Sessions 1–3)

- Scissors: 1–2 per small group (Sessions 1–3)

- Colored markers: several per small group (Sessions 1–3)

- Newspapers: 10 editions of the same paper, one from each of 10 different days. *USA Today* is easy to work with because there are no Sunday editions and no extra sections devoted to ads. If you prefer to work with your local newspaper, call the marketing department and ask what percentage of the paper is their goal for ads. (Session 2)

- Adding machine tape: about 40 feet (Session 3)

- Glue stick with a removable adhesive, or removable tape (Session 3)

Other Preparation

- Gather some old newspapers for practice work. Each of 10 groups will need one section of a newspaper, preferably one that has a good mixture of ads and writing. Have a few extra sections for homework as needed. (Session 1)

- To get a random sample of newspaper pages, prepare a bag with numbered slips corresponding to the pages. (Session 2, optional; see p. 89 for other options)

- Duplicate Recording Strips (p. 151): 3–4 sheets per small group. (Session 2)

- Using two sheets of the Recording Strips (p. 151), color a set of 10 strips with different fractions, as if you had taken data from 10 newspaper pages (see p. 91, for example). Cut a 30-inch length of adding machine tape to demonstrate to the class how they will combine their strips. (Session 3)

Fractions of Newspaper Pages

Materials

- Sections of old newspapers (1 per small group, plus extras for homework)

- Scissors (1 per small group)

- Colored markers (1 per small group)

- Rulers, tape measures (1 per small group)

What Happens

Students practice finding the fraction of a newspaper page that is made up of ads, using a variety of strategies. They describe and compare their strategies. Student work focuses on:

- estimating fractions of areas
- combining and comparing fractions represented as areas

Activity

Teacher Checkpoint

How Much of the Page Is Ads?

Introducing the Problem Explain to students that they are now going to work on a different kind of sampling problem.

We're going to study one way a newspaper might use samples. Do you know how publishers get the money they need to publish a newspaper? What people pay for their paper is only a small part of the money a newspaper needs. Most of the money comes from selling space for ads. This is especially true with the newspaper *USA Today,* because many people get their copies free.

Tell students that at *USA Today,* the goal is to sell ads to fill two-fifths of the paper. (This information was provided in August 1994 by Stew Bradley, Boston Advertising Sales Manager of *USA Today*.)

We're going to look at a sample of issues of *USA Today* to find out whether they achieve their goal of two-fifths ads. I've collected a sample of ten newspapers. We'll figure out what fraction of these newspapers is ads. What are some ways of doing this?

[Hold up a page with some ads.] **If we just wanted to figure out what fraction of this page is ads, how could we do it?**

Students will probably have several suggestions. They might want to count the columns, then figure out how many column-equivalents are ads; they may want to cut out the ads from a page, combine them, and see how much area they cover; or they may suggest some other measuring strategy.

Form groups of two or three students. (They will work in the same groups in the next session.) Ten groups is convenient, and we will refer to ten groups in this Investigation, but if you have too few students, you can reduce the number of groups to seven or eight. There must be at least two students in a group.

Practice with Newspaper Pages   Before students work with the series of *USA Today* newspapers, they try out their data collection strategies with a practice page from any paper. Give each group a section of a newspaper to work with, as well as colored markers, scissors, and measuring materials.

Before you figure out what fraction of your page is ads, we need to decide what will count as an ad and what won't count.

Discuss what will count as an ad in your data collection. For example, Are classified ads, movie announcements, and legal notices counted as ads? More examples are likely to come up as students actually begin to look at the newspaper, and the class may have to make a few more decisions as they examine individual pages. It is important that these decisions are made during this first practice session, so students' data collection in the next session will be consistent.

First your group should choose a page that has some ads, but that isn't all ads. Then use any strategy you want to figure out the fraction of ads on the page. Try at least a couple of ways, so you can see which works best. Work with familiar fractions like tenths, sixths, eighths, fourths, thirds, and other fractions you can easily figure out. Your job is to get a good estimate, not an exact number.

Encourage students to work on pages that aren't obviously one-half or one-fourth ads, or some other easy-to-see proportion.

What to Look For   Your role during this time is to listen and help students articulate their strategies for determining what fraction of the page is ads. At the same time, you will want to make sure that students develop good estimation and measuring strategies before they go on to collect their *USA Today* data. Look for the following:

- Do students have good strategies for dividing the page into parts? Do they make use of folding, measuring, cutting and rearranging, or counting column-equivalents?

- Do students use their knowledge of familiar fractions in doing this task?

- Are students able to find a familiar fraction that is close to an unfamiliar fraction, for example, to see that $14/36$, a fraction they might get from counting column inches, is close to $1/3$?

- How do students express and combine fractions that are not alike, such as $1/6 + 1/2$ of $1/6$?

Expect to see a variety of strategies for figuring out fractions. Some students may cut up the page and rearrange it like puzzle parts to figure out what fraction is ads. Others may use markers to color the ads and visualize the fraction of area they take up. Still others may use a counting strategy, making use of columns or even column inches. (Many newspapers sell ads by column inch; *USA Today* sells ads by fraction of a page.)

At the end of the session, groups share their strategies for figuring out fractions of the page. List the strategies on the board for students to refer to. During the next class session, each group will choose any strategies they like—not necessarily the ones they used today—and use them to find the fraction of ads on a sample of pages of a newspaper for a given day.

Session 1 Follow-Up

Homework

Students continue the Session 1 activity at home with their own newspaper, choosing one page and reporting the fraction that is ads. (Have a few extra newspaper sections available for students who do not receive a paper regularly at home.) Be sure students know not to choose a page that has no ads, is half ads, or is all ads.

Their report should include the following information (they should write this down to take home with them):

1. name of the paper
2. the page number
3. fraction of the page that is ads
4. fraction of the page that is not ads
5. an explanation of how they figured out the fractions, using words or pictures or both

Collecting Data from Ten Days

What Happens

Students define a sampling strategy so they can take data from their newspaper in a single class session. Each group figures out ad proportions on a sample of pages in a single day's newspaper and records the data. Student work focuses on:

■ defining and carrying out a sampling strategy

■ expressing fractions of areas as familiar fractions

■ organizing a complex data-collection process

 Ten-Minute Math: Volume and Surface Area During the next few days, continue to do this activity with cube configurations from the transparency master (p. 154). For example:

How many stamps will it take to cover this solid? How many cubes make up the solid?

Materials

■ Recording Strips (3–4 sheets per small group)

■ 10 days' issues (1 per group) of *USA Today*, or another paper

■ Scissors (1–2 per small group)

■ Colored markers (several per small group)

■ Rulers or tape measures (1 per small group)

■ Bag of numbered slips for random sampling (optional)

Activity

Begin by asking students if they changed their strategies for finding fractions of a page or came up with better strategies when they did the homework. Remind students that today their groups will find the fraction of ads on a sample of pages of their newspaper.

You have some good strategies now for figuring out what fraction of a page is ads. Some of the pages you'll examine will be pretty easy. Some pages will be harder. For some pages, you could get fractions like 3/28. What familiar fraction might you use instead if you got this fraction?

Sampling and Collecting Ad Data

Introducing the Recording Strips Distribute the sheets of Recording Strips to the small groups and describe how to use the strips.

Each of these sheets gives you six Recording Strips. For each page of the newspaper that you examine, you'll need to record what fraction of that page is ads. Use a separate Recording Strip for each page. On the strip, color the fraction of the page that is ads. If a page has no ads, you still need a strip for it; just leave the whole strip uncolored. If a page is *all* ads, you would color the whole strip. What would you color if a page has one-eighth ads?

Students write the page number on each strip they use so they can keep track of which pages they have looked at. When students are clear on this procedure, show them how to set up a data sheet on notebook paper. Here each group will record the page number, section, and fraction of ads for each page they examine, in addition to coloring the Recording Strip.

They label a line for the day and date of their newspaper at the top of the page, and leave several lines to describe their sampling plan.

Day and date of newspaper:		
Sampling plan:		
Page number	Section	Fraction of ads

Choosing a Sample Now turn the discussion to sampling.

I've collected ten copies of *USA Today* for you to analyze. Each group will get a whole newspaper, with all the sections, and your job is to figure out what fraction of the paper is ads.

You won't have enough time to do this for every page, so you need to choose a sample of pages from the paper. You will have time to figure out the fraction of ads for 10 to 15 pages of your paper. What are some reasonable ways of sampling the paper? How will you get a representative sample?

Discuss students' suggestions for taking a reasonable sample. During this discussion, students are likely to comment on the fact that some pages of the paper generally have more ads than others. See the **Dialogue Box,** Sampling from the Newspaper (p. 90), for the kinds of suggestions you might hear. Some sampling methods that will work are as follows:

- Choosing pages randomly. Pick page numbers out of a bag or use a random-number program on a computer or calculator.

- Counting every third, fourth or fifth page, depending on how many pages the newspaper has, to sample 10–15 pages. For example, if the newspaper has 37 pages, use every third page.

- Sampling by section. Decide how many pages should be counted in each section, depending on how long that section is. Then choose the actual pages by either method above.

After this discussion, each group chooses its own sampling strategy and writes it on their data sheet.

Collecting the Ad Data Distribute the copies of *USA Today*, one per small group. Students work on their newspaper for the rest of the period. As they work, observe the different groups. Ask questions like those suggested in the Teacher Checkpoint in Session 1 (p. 84) to help students focus on the problem of finding a good estimate of the fraction of ads on each page they examine. If necessary, help groups organize their data collection and recording so that they can complete at least 10 and preferably 15 pages. If some groups finish before the end of the period, they can help other groups by taking data on some of their pages.

Session 2 Follow-Up

If students have not finished working on their sample in class, they divide up the remaining pages among group members and take them home to finish.

 **Homework**

Sampling from the Newspaper

During the activity Sampling and Collecting Ad Data (p. 87), these students are discussing possible strategies for choosing a representative sample of 10–15 pages from their newspaper.

Antonio: We could like start with the first page, and do every other page until we get 15.

Heather: How long are newspapers? Maybe we wouldn't get enough pages.

Look at the last page of your newspaper and tell me the total number of pages in it.

Heather: Forty-eight.

Shakita: Forty.

Tai: Forty-eight.

Jeff: Thirty-six. I think the problem would be that if you did every other one, you'd only cover 30 pages. And so if your newspaper was longer, you wouldn't get any of that section.

Tai: Maybe we should just take one section and do all of the pages in it.

Heather: But the sections are different! Look, there's lots of ads in this section, lots of car ads especially. I think sports has the least ads.

You're saying we shouldn't just choose one section?

Heather: Yeah, because they're not the same. You should try to be more fair in your sample.

How could we sample more fairly?

Shakita: How about if we took every third page? Then if it were 36 pages long, we'd have a sample of 12. If it's 48 pages, we'd have a sample of... Does someone have a calculator?

Tai: *[Entering 48 ÷ 3 on a calculator]* It would be 16. That's pretty close to 15. It wouldn't matter so much if we didn't do the last page.

How does that seem to the rest of you?

Heather: I think it will work most of the time, unless we find a really long newspaper or a really short one.

Combining Ad Data Across Pages

What Happens

Each small group combines their data to calculate a single fraction for the entire paper. Then the groups compare their results and discuss their sampling strategies. Students compare their data to the newspaper's stated goal for ads. Their work focuses on:

- totaling a set of fractions using a visual model
- comparing fractions represented visually and as numbers
- comparing sampling strategies

Materials

- Adding machine tape (30-inch length) and 10 colored-in Recording Strips (demonstration)
- Each group's completed Recording Strips
- Adding machine tape (about 40 feet)
- Tape measure
- Scissors and rulers (1 per small group)
- Glue stick or tape
- Colored markers (1 per small group)

Activity

Today, each group will put together their Recording Strips to figure out what fraction of their sample of the newspaper is ads. Here's the main question we're asking: If all the ads in your sample were grouped together, how many pages would they fill up? What fraction of the pages in your sample would that be?

Explain that while there are many different ways of finding the combined fraction, everyone in the class will use the same method so that it will be easier to compare the findings. Demonstrate this method, taping your 30-inch piece of adding machine tape to the board. Then fasten your ten colored-in Recording Strips to the adding machine tape so that they cover it. Glue stick with a removable adhesive or removable tape is best, since you will be removing the Recording Strips, cutting them, and retaping them.

Putting the Fractions Together

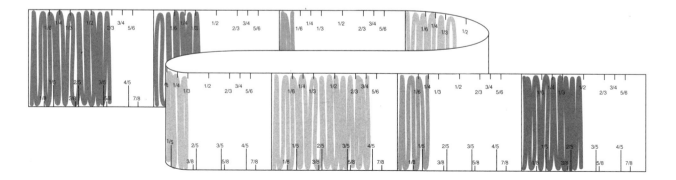

What does this whole paper tape now represent? (all the sampled pages added together) **How might we use these Recording Strips to figure out what fraction of these pages is covered with ads?**

Listen to students' ideas on how they might manipulate the strips to represent the fraction of ads in these pages. The idea is that the colored-in part and the blank part of each strip need to be cut apart so the colored parts can be grouped together. Take the Recording Strips down and cut off the colored-in part of each one. Save the blank pieces to show later how they fill in the rest of the strip. Tape the colored-in pieces onto the paper tape, starting from one end of the tape and putting the pieces right next to each other (see example at bottom of page).

What do the remaining blank parts of the Recording Strips represent? If I put them next to one another, they will cover the rest of the strip that isn't covered by the colored-in parts. Why?

How could you now use this tape to figure out what fraction of my newspaper sample was ads?

Students should see the similarity of this task to the work they did earlier in this unit with Data Strips. In each case, part of a strip represents one response to a question, the whole strip represents everything or everyone surveyed, and they can use folding, numerical reasoning, or calculating to find a familiar fraction that is close to the fraction on the strip.

Now students work together to combine their group's Recording Strips. Give each group a piece of adding machine tape that is 3 inches times the number of pages they have sampled. For example, if they have sampled 12 pages, give them a 36-inch length of adding machine tape. The group cuts each of their Recording Strips into two pieces and fastens down the colored-in parts, starting at one end of the tape.

As the groups finish, ask them to write the day of the week, the date, and the fraction of ads in their sample on the blank part of their strip. They can fold their strips to figure out this fraction, or use any other strategy that makes sense to them. To use the same technique they used with Data Strips, students may make another strip to fold and use as a fraction strip to compare with their combined Recording Strip.

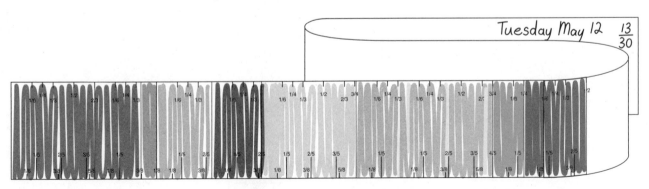

Note: The way fractions are combined in this procedure is usually easy for students to grasp, but the underlying mathematics is a bit tricky. Be sure that students don't look at the strips they are adding together and decide that ½ + ½ = ½ or 2⁄4, even though they know from their work with fractions that ½ + ½ = 1. They may make this mistake because when they put together two half-shaded Recording Strips, they get a larger strip that is half-shaded, as shown here.

  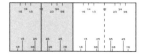

In fact, the original two halves are halves of the smaller strip, but the new half is half of the larger strip, which is twice as large as the smaller strip. If we put together just the shaded portions of the two smaller strips and discarded the unshaded parts, we would get our expected answer of 1, as shown here.

Watch to be sure that no one draws the wrong conclusion from this procedure.

As students finish their paper tapes, post the tapes on the wall next to each other to make it easy to compare them (although the paper tapes will be different lengths if groups have sampled different numbers of pages). Each group labels its tape with the day of the week they studied and the final fraction of ads they found. Students might express their fractions many different ways, from 15⁄32 to "3⁄8 and ¼ of 12." Discuss the different expressions and compare them to the fraction represented by the strip.

Groups that finish early can make a second paper tape the same length as the one they just made. Ask them to color in two-fifths of this tape, so they can see how the fraction of their sample compares to the two-fifths figure that *USA Today* is striving for.

Give students a little time to examine the paper tapes that they've posted. Then begin a discussion.

How similar or different are the findings from the different days' newspapers? Is there about the same fraction of ads each day?

Drawing Conclusions from the Sample

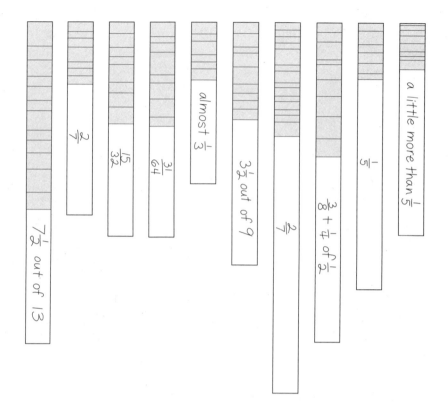

As this discussion proceeds, observe how students are comparing the fractions represented by the tapes. Some students will want to simply compare the size of the area colored in on each tape. Probe their thinking on this a bit more:

Does more area colored in always mean a bigger fraction? Why not?

You may want to compare a short tape with a longer tape, especially if they have roughly the same fraction of ads. Help students focus on the relationship between the colored-in segment and the entire strip for each tape, rather than the comparing the size of the colored-in segments. You might refer back to comparing samples in your class with the entire class, or comparing your class with the larger sample of students across the country in the Meals and Chores Survey.

Put the paper tapes in order of the days of the week and ask students to look for a pattern.

Does there seem to be a pattern in the fractions of ads on different days? Are there more ads early or late in the week?

Now ask the students to consider what their findings tell them about the original question.

Based on our sample, do you think *USA Today* is reaching its goal of two-fifths ads? What would you conclude? Are most of your figures above, below, or right around two-fifths? Is there much difference between issues?

Finally, focus on the effect of different sampling techniques and students' general observations of how ads vary throughout the paper.

Do you think your data are representative of *USA Today* in general? Why or why not? Did we take data from enough days? If you were to do this again, how might you change your sampling plan?

Session 3 Follow-Up

 **Extensions**

- **One Long Data Tape**   After looking at the individual tapes, students might cut them apart and recombine them into one large tape, which can be displayed along a wall or in the hallway. How could they find out from this tape what the total fraction of ads in their 10-day sample was?

- **Ads by Section**   If you have students for whom this work is fairly easy, suggest that they investigate how ads are distributed by section. Ask students what they notice about the similarities and differences in the fractions of ads in the different sections.

  Are there certain sections that always have more ads than others? Why do you think the publishers of the newspaper make different decisions about the fraction of ads in different sections? What can you learn from the data we have about the fraction of ads in each section?

  Make copies of all the student data sheets from the original activity, and give each small group a complete set of data sheets. Groups reorganize the data by section and determine the fraction of ads in each section. Since combining proportions of ads for each section will require adding fractions, students will probably want to color more Recording Strips, cut them apart, and combine them on adding machine tape.

- **Types of Ads**   If you are using *USA Today,* students might look at the full- and half-page ads and compare them to ads in a local paper. Are there differences in the types or percentages of ads in corresponding sections of the two papers? Why might this be so?

- **Other Newspapers**   Students bring in other newspapers to compare to the one they have been studying. How do community newspapers compare to national ones? How do foreign-language newspapers compare?

- **Magazines Versus Newspapers**   Students might look at the fractions and types of ads in magazines and compare them to newspapers. They could also compare the fractions and types of ads in different kinds of magazines.

Researching Play Injuries

What Happens

Session 1: Issues of Playground Safety
Students consider how safe their school playground is, and learn what other researchers have found about playground injuries. They formulate survey questions about injuries and try out these questions as homework.

Session 2: Collecting Playground Data Based on their preliminary data collection, students refine their questions and develop a sampling plan for gathering data about injuries on the playground. Between Session 2 and Session 3, students collect data.

Sessions 3, 4, and 5: Analyzing and Presenting Data Students combine their data into a single database. Each group investigates a question and draws conclusions using their own charts, tables, and graphs. If database software such as Tabletop is available, students use it to investigate questions about the data. As an assessment, students present their findings in poster form and discuss them. The entire class considers what recommendations to make about playground safety, based on their conclusions.

Mathematical Emphasis

- Defining the question and the sample
- Testing and refining survey questions
- Collecting and collating data
- Deciding on appropriate representations for data
- Looking for associations and developing theories based on data
- Making recommendations based on interpretation of the data

What to Plan Ahead of Time

Materials

- Chart paper (Sessions 1–5)
- Clipboards (optional, for data gathering between Sessions 2 and 3)
- Calculators: 2 per small group (Sessions 3–5)
- Construction paper, scissors, colored markers (Sessions 3–5)
- Database software, computer (optional, Sessions 3–5)

Other Preparation

- For Session 1, duplicate Student Sheet 12, Danger on the Playground (p. 152), 1 per student.
- After Session 1, you will need to make copies of the students' survey questions, one for each survey subject, for homework.
- After Session 2, students will be conducting a survey and bringing their results to class. There are several options for collating this data, as described at the beginning of Session 3 (p. 110). After the collation process, plan on making a copy of all the data for each small group to work with.
- Before you start this investigation, you may want to read some details of one classroom's work on this topic from start to finish. The **Dialogue Box,** A Sample Playground Study (p. 104), follows a combination fifth-sixth grade class's work on playground injuries, including their presentation of their results to a group of parents of students at the school.

Session 1

Issues of
Playground Safety

Materials

- Student Sheet 12
  (1 per student)
- Chart paper

What Happens

Students consider how safe their school playground is, and learn what other researchers have found about playground injuries. They formulate survey questions about injuries and try out these questions as homework. Student work focuses on:

- formulating survey questions
- using background information in designing a survey

Considering the Problem

Note: The data gathering for this investigation is easiest if students collect data about injuries that happened on their own school playground. If there is no playground at your school, choose a nearby one to work with, or study injuries within the school building itself. The discussion here will refer to injuries on the school playground; adapt it as needed to your situation.

In this investigation, your challenge is to become researchers studying one problem in depth. The problem is injuries on our playground, and you will collect data by asking other students about injuries they have had on the playground.

Let's start by thinking about ways you have been hurt while on the playground. What were you doing when the injury happened? What do you think caused the injury?

These questions will provoke some lively discussion. Fifth and sixth graders love to trade gory stories about their injuries. For some possible ways of handling these issues, see the **Teacher Note,** Sensational Issues in the Classroom (p. 103). As students discuss their injuries, make a list of the types of injuries and the circumstances surrounding them.

As the discussion proceeds, ask students about injuries they have witnessed on the playground, including injuries to younger children. Since students will have to agree on what counts as an injury for their survey, an important goal of this discussion will be to move toward a decision on this topic. See the **Dialogue Box,** What Counts as an Injury? (p. 106).

After students have discussed their own and other students' playground injuries, distribute Student Sheet 12, Danger on the Playground. Tell students that a group of researchers did a survey about things that might be dangerous on playgrounds. Encourage students to think about whether any of the danger factors described might be present on their own playground.

❖ **Tip for the Linguistically Diverse Classroom** Use sketches and simulated enactments to make the information on this sheet comprehensible to students with limited English proficiency.

When students have read the page, have a brief discussion of the findings. What was surprising to students? How do the findings relate to their own experiences and their hunches about playground injuries?

A Note on Organizing the Data It is important to establish a clear and consistent idea of the structure of the data students will be collecting from the beginning of the investigation. Students could carry out many kinds of surveys to investigate playground injuries, but some would lead to uninteresting data or would present special difficulties with data collection and representation. This investigation is organized around a particular data-collection plan that will help avoid some of the most common pitfalls of student surveys. You can decide how much of this structure to make explicit at the beginning of the investigation and how much to allow to emerge through the discussion.

The data students collect will be about injuries, so each completed interview will be about an injury, not about a person. When all the students' data are put together, they will look something like this (but with your class's choice of column headings).

Injury	Severity	Age when injured	Where injured	Day of week	Other data
bloody nose	medium	7	Sliding board	Wed.	
broken arm	high	5	jungle gym	Sat.	

❖ **Tip for the Linguistically Diverse Classroom** In a chart of this nature, a symbolic rating system could be used to clarify the severity column. For example: High • • • • •, Medium • • •, Low •.

In order to look for relationships between variables, each small group will need a copy of the entire data set. Limiting the number of questions asked and injuries recorded will make the logistics of this process more manageable. About eight to ten questions and about 50 to 60 injuries in the database will be workable in most classrooms. Note that 60 injuries, with 8 questions about each, adds up to 480 answers. Several suggestions on the logistics of combining students' surveys to facilitate analysis are included in the activity Compiling the Data (p. 110).

Activity

Defining the Survey Questions

Now that you have talked about your own experiences and seen some national research on playground injuries, it's time to think about questions we might ask to get some data from other students in our school. We want to look for patterns in the injuries on our playground.

We will do this by interviewing people who use the playground, recording the injuries they report, and looking for patterns in the information. Let's come up with a few questions we'd like to find answers for.

Students may have some theories they want to test: a particular location they want to focus on, a specific kind of equipment, or a specific need for safety equipment on their playground. They may have questions based on their own experiences, on the experiences of younger children, or on findings that interested them in the research study they read about.

As students identify the issues they want to study, make a list on chart paper. In one class, the list looked like this:

Issues to Study

Do younger children get more injuries than older ones?

Do kids get injured more on equipment or during games?

What kinds of injuries do kids get at recess?

Do younger children have more injuries after school than they do during school? How does this compare with older children?

Do fifth and sixth graders have more injuries during the week or on the weekend?

Do younger students have fewer injuries when a teacher is around? Is this true for older students too?

Save the students' initial list of Issues to Study for reference when they analyze their data later in the project (Sessions 3–5).

Once students have raised some general concerns, they need to decide what data they will collect and how they will word their questions. Although the main source of their data should be interviews, they may want to augment these with information from the school nurse's records and/or their own observations of behavior on the playground.

Now that you know what you want to look at, there are two questions to answer:

1. What data will you need to collect?

2. What questions will you ask people in your survey?

Before working on the wording of the questions, students make a list of the information they need to collect, such as the kind of injury, what the injured person was doing, where and when the injury happened, the age and gender of the injured person, and so on. Based on this list, students make up specific questions to elicit the information.

For example, students might decide on questions like these:

What was your injury?
What were you doing when you were injured?
Where were you on the playground?
When during the day were you injured?
When during the week were you injured?
How old are you?

Record proposed questions as students talk, shaping and reshaping their questions as they suggest more ideas. You may immediately recognize that some of their questions are problematic. It's very tempting to simplify students' questions to save them from wrestling with messy data or overwhelming amounts of information. But struggling with these issues is an important part of the process of working with data. Students will have a chance to try out their questions and learn for themselves how they work.

Remind students that they will have to agree on what they will consider an injury *before* they go out to interview. That way, they can decide quickly whether or not to ask an interviewee all the questions. If the person has not had an injury, the student needs to go on to the next person. Be sure students have made a substantial start on the definition of an injury before the end of this session, so they can try out their first ideas for homework.

At this point in the planning, students often suggest questions such as, "Do you think the playground is safe?" Help them stay focused on the subject by reminding them that in this study, they are trying to find out about injuries, not about other students' opinions. Some students may want to carry out a supplementary study on other students' opinions about playground injuries, but it should be kept separate for now.

When students have agreed on the questions they want to ask for their survey, either photocopy a sheet with all the questions or ask students to write them down to take home.

Session 1 Follow-Up

 **Homework**

Students try out their survey questions at home before using them to collect data. They get someone to interview them at home, so they can see what it's like to answer the questions and how well their questions work. They can also interview siblings or neighbors to find out more about how the questions work. They write their assessment of the questions and revise them as they see fit, based on their trial.

❖ **Tip for the Linguistically Diverse Classroom** Students with limited English proficiency can conduct this trial survey in their primary language.

Sensational Issues in the Classroom Teacher Note

Ten- and eleven-year-olds have always loved being graphic about horrible accidents that have befallen them. If you ask them to talk about play injuries, they are likely to fling themselves into a "can-you-top-this" kind of discussion.

One of the reasons for their interest is that many students this age do not believe that they can be seriously injured. They feel invulnerable. On the other hand, they know intellectually that terrible things happen. They may live in a neighborhood that has serious crime; they may know someone who has lost a family member to violence. They may have experienced the loss of a family member or friend through an accident. Typically, these early adolescents are fascinated with accidents and the details of them.

Students at this age want to develop a sense of what to do if there's a serious accident. They enjoy learning about first aid treatments and want very much to be able to cope in an emergency. They also like to know that they are getting old enough to help protect others. Doing research about a safety issue puts them in the front lines, protecting both themselves and others from risk.

Because of their need to feel effective, fifth graders enjoy studying a problem that has real-life consequences and that they might help solve. Studying play injuries can lead to real changes on the playground, in the park, or at home. If these changes can be geared to protecting younger children, so much the better.

The "blood and guts" kind of discussion can be a lead-in to this concern on the part of students. The teacher's role in such a discussion is delicate. You can certainly focus on the safety aspects of the accidents, asking questions like these:

> What time of day was it?
>
> How did you get help in taking care of yourself?
>
> How did you get to the hospital?

In some instances, you may prefer to provide the reassurance that students of this age still need from adults, with comments and questions like these:

> That's not very likely to occur again. How unusual that it happened.
>
> Bleeding like that is frightening. It's a good thing you stopped it by pressing on the cut.
>
> I'm sure your sister is more careful on the slide since her fall.

At a certain point, you can turn the discussion away from gore and toward data collection:

> How could you record that to add it to our data?
>
> Should you count only cuts that need stitches, like yours did?
>
> What would Rachel write if she were interviewing you about that accident?

Playground safety data present a unique opportunity for your students to take an active role in protecting themselves and others. You have an unusual chance to help your students see that research and recommendations based on statistics can be a way to take action in a difficult, complex environment.

Session 1: Issues of Playground Safety ▪ **103**

◻D◻I◻A◻L◻O◻G◻U◻E◻ ◻B◻O◻X

A Sample Playground Study

This dialogue box follows a class through their playground study, from composing the questions to reporting their findings to the school board. The brief segments provide a glimpse of what to expect in this project.

The first step in the project (Session 1, p. 100) involves choosing questions to ask in a survey—questions that will give them the data they need for their study.

What's the big point of the survey?

Robby: To figure out what percent got hurt and how.

Amir: And also where and what they were doing.

Katrina: I think we ought to first ask about whether they ever had an injury that was serious enough to go to the nurse's office or the emergency room.

Cara: There could be three parts in a question: First, have you ever gotten injured on the playground? Then, did you ever have to go to the nurse's office? And then, did you ever have to go to the emergency room or hospital?

Christine: Then we'd get an idea of how serious it is, because some kids will say it's an injury even if it's just a scratch.

What else should we ask?

Desiree: We need to know where they got injured. That should be simple.

Robby: Just make sure to ask, "Where on the *playground* did you get injured, so they don't think it's where on your body.

Desiree: Let's also ask what they were doing when they got hurt.

Amir: I think we need to ask how they got hurt.

Duc: Isn't that the same as what they were doing?

Amir: No, because they might have gotten hurt by falling, but what they were doing was trying a gymnastics trick. In the real survey, they wanted to know how many injuries were caused by falls or being hit by equipment.

After selecting and refining the questions they are going to ask, students address the need for sampling and talk about ways of getting a good sample (Session 2, p. 108).

We have our questions; now we need to know how to select a good sample. We can't have more than 100 kids, because it would take too much time.

Mei-Ling: Well, we should get some younger and some older kids. How many kids are in a class?

About 25 or so.

Mei-Ling: What if we took four classes? We could do ourselves…

Matt: I don't want to do us, we're boring!

Mei-Ling: Well, then the other fifth grade, and maybe a first grade, a third grade, and a seventh grade.

Antonio: How about if we do some kids from each class? That way we wouldn't leave out a class.

Matt: Yeah! We could put all the kids' names in a hat for each class, and draw out some. That way, they'd feel special if they got picked.

How many would we choose from each class? There are 18 classes in the school, two at each grade level.

Desiree: If you did four per class, it would be 18 × 4 *[using a calculator]*…that's 72, so that's not enough. Five per class would be about right, but it's still a little short.

Continued on next page

Matt: I think we should do an equal number of boys and girls, and we can't get that if we do five. Could we do six, and have three of each?

How many would that be altogether?

Katrina: 18 × 6 = 60 and 48, which is 108. That's close enough. Besides, someone might be absent.

With a sampling plan defined, students proceed with their survey and compile their data. In analyzing the data they have collected, other issues arise, such as finding a way to categorize the data and look for associations. The **Dialogue Box,** Categories and Associations (p. 119), offers examples of student thinking at this stage.

Finally, the students appear before the school's parent advisory board to report their findings and recommendations.

Parent: How did you decide what counts as a serious injury?

Greg: Well, it was really up to the kid to decide. But we also asked them whether they had to go to the emergency room.

Becky: If they did, it was definitely serious.

Yu-Wei: And if they just went to the nurse's office, it was medium.

Parent: You said that the bars were pretty dangerous. What could we do to make them safer?

Leon: Most of the injuries on the bars were from falling. "Trip or fall" was our biggest category of injuries. So we could do something to make sure that the ground's softer.

Christine: Maybe put more mulch there.

Robby: Or not let kids play on the bars in winter. Because the ground's frozen and the bars could have ice on them. A couple of kids got hurt because they slipped when there was ice on the bars.

Parent: Did you think at all about the playground rules and what they have to do with safety? Should we have different rules?

Manuel: Well, a lot of kids got hurt when they were running. We could make a "no running" rule, but that would sort of take the fun out of it.

Rachel: I think it's more important to have rules about the number of kids in an area. Because the more crowded it was, the more kids seemed to get hurt.

Jasmine: We could make rules about how long each group could be on a piece of equipment. Maybe they could take turns.

Parent: What differences did you find between younger and older kids?

Shakita: Mostly, kindergartners were hurt more in slipping and tripping injuries.

Robby: Older kids got hurt more in fights, and in falling from bars. Actually, a lot of kids of all ages got hurt falling from bars.

Principal: The next step here is to take these findings to a faculty meeting, and see what we can do to make the playground safer.

What Counts as an Injury?

These students are beginning work on their study of playground safety. They are trying to decide how to define an injury (p. 100) before they begin their survey.

What injuries have you gotten on the playground, and how did you get them?

Amy Lynn: Remember last year when I broke my arm? That's because I was hanging upside down on the bars and when I tried to come down, one leg got stuck. I landed wrong. Right on my shoulder.

Katrina: Yeah, you were really crying. And you were in a cast for a long time.

[The teacher writes "broken arm—fell from bars."]

Tai *[excitedly]*: I've got one! You know that old slide on the side of the school? I was going down it backwards and I was going too fast when I got to the bottom. I fell really hard and had to go to the doctor. She said I bruised my tailbone.

Maricel: That slide is really dangerous, even when you're going forward. When I was little, I got hurt from falling when I was just sliding the regular way.

[The teacher records this information on the list.]

Kevin: I got a bruise too, when we were playing Ships Across the Ocean and Shakita grabbed onto me and pulled. My arm was black and blue for a week.

Maricel: Yeah, but that's not a real injury. There was no blood or anything.

[The teacher writes down Kevin's information.]
What do you mean by "real injury"?

Amy Lynn: I don't think there has to be blood, because I didn't bleed when I broke my arm. But I was really hurt.

Sofia: Amy Lynn's injury was real. And even though Tai didn't look hurt when he fell off the slide, he had to go to the doctor. So that was a bad injury too. But I don't know about Kevin's.

Kevin: Black and blue counts. Bruises count.

Sofia: What about if you just get a scratch and it bleeds a little? That happened to me when we were playing kickball.

Amy Lynn: If it bleeds and you have to go to the office, it's a real injury. And also if you have to go to the doctor. If you just keep playing, I don't think it counts.

[The teacher writes this definition of an injury and the class considers it and other definitions for a while. A few minutes later, the teacher summarizes the kinds of injuries they have listed so far.]

Are there some kinds of injuries that you've seen happen to little kids, that we haven't listed yet?

Danny: They play jump rope a lot, and I think they get bloody knees more than we do. Especially girls who wear skirts.

Shakita: Yeah, and they get in fights more often because they don't know the rules yet. They push more, and sometimes they hurt each other.

How should I list that?

Julie: Write "falling down injuries from fighting."

Tai: But it has to be serious enough for them to go to the office. A lot of them cry, but that doesn't mean it's serious.

Tai is getting back to the definition of injury again. Does this kind of injury fit in with what we decided earlier?

Julie: I think it does, as long as we change the part about "going to the office because you're bleeding" to just "going to the office because you're hurt."

Tai: It's not enough just to cry. An injury is more serious.

You have a good start at defining what you mean by "injury."

Collecting Playground Data

What Happens

Based on their preliminary data collection, students refine their questions and develop a sampling plan for gathering data about injuries on the playground. Between Session 2 and Session 3, students collect data. Their work focuses on:

■ defining and carrying out a sampling plan

■ organizing and carrying out the process of data collection

Materials

■ Chart paper
■ Clipboards (optional)

Refining the Questions

You tried out our survey questions last night as homework. Did anyone run into problems with the questions? Were there any surprises? Do you need to change any of the questions?

Allow some time for students to talk about their experiences with the questions. Encourage them to use what they wrote for homework to make suggestions about how the survey questions could be improved. Make a list of the revised questions on chart paper or the board, and keep working with students until they are satisfied that the questions are clear and useful.

If students wish to add or delete any questions, now is the time to do so. Aim for eight to ten questions (including gender and age) on the questionnaire; more than this will make it difficult to collect and analyze data.

Coming to closure on a set of questions and organizing the data collection are challenging tasks. If they are given the opportunity, most classes can spend several sessions in discussion before they come to a conclusion. Guide your students to conclusions about the survey questions and the sampling plan during this session; their time is better spent on analyzing the data (Sessions 3–5).

Choosing a Sample

Think about where to go from here. We have our questions, and now we have to decide who our sample will be. We also need to figure out a plan for gathering data. Any ideas?

The challenge is to come up with a plan that will yield information on an average of three injuries for each interviewer (to keep the total amount of data reasonable), and will have enough variation in age in the sample to allow students to explore age differences in injuries.

In some classrooms, students have found at least half and even as many as 80 percent of subjects reported some kind of injury on the playground (although not necessarily one that they would count as serious enough to report). If this pattern holds in your school, it would be best for students to interview about 100 subjects to get data on 50–60 injuries.

One plan would be to include your class and two to four other classes—say from first and third grades—in the interviews. Each student could interview one person in each class, or small teams of students could go into each of the other classes. You would then have sampled three age groups and should have about the right amount of data. This method—like any method in which you know how many people were interviewed in total—has the advantage of allowing students to calculate the proportion of people in each class that had injuries.

Various factors could affect your sampling decisions. If you have a computer with a database program, it is less important that you limit the amount of data your students collect, since some students can enter the data and then make multiple copies of the data set. If you have no automatic way to collate the data and cannot afford the extra day it might take to do it by hand, you may want to limit data collection even further.

Here are some things to consider in the discussion (you may want to decide some of these yourself):

- Which age groups will we sample—all the age groups in the school or just certain ages?
- Should we choose a few classes and ask questions of everyone in those classes, or choose a few students from each class?
- How many students do we need to interview?
- Where and when should we ask the questions?
- How can we be sure that our sample is representative?

Remind your students that they will record data only for those interviewees who have had an injury, by the class definition. This means that they won't be able to predict ahead of time how many pieces of data they each will gather. But, if students keep track of how many people they each interview, they will have the data to calculate what fraction of the sample reported injuries, and whether this fraction changes with the age of the interviewees. Be sure that students organize their interviews so that they don't talk to the same person more than once.

Once students have decided on a sample and have a plan for when and how to gather data, do some role playing for the rest of the class period. Students may want to work in pairs and practice interviewing each other. Alternatively, you might role-play an interview in front of the whole class.

A note on interviewing etiquette: Remind students that it's all right for someone to refuse to be interviewed, and that they should respect that person's wishes.

Write students' final list of questions on a sheet of paper and make enough copies for students to interview everyone in their sample. Schedule the next class session for after the data are collected.

Session 2 Follow-Up

If the survey subjects are not students in other classes, but rather people in the broader community, the interviews can be conducted as homework. Students take copies of their sheet of survey questions and interview people according to their sampling plan.

Homework

Analyzing and Presenting Data

Materials

- Students' interview results
- Class list of Issues to Study (from Session 1)
- Calculators (2 per small group)
- Chart paper
- Construction paper, scissors, colored markers
- Database software and computer (optional)

What Happens

Students combine their data into a single database. Each group investigates a question and draws conclusions using their own charts, tables, and graphs. If database software such as Tabletop is available, students use it to investigate questions about the data. As an assessment, students present their findings in poster form. The entire class considers what recommendations to make about playground safety, based on their conclusions. Students' work focuses on:

- organizing and representing data
- deciding what are the most important features of complex data
- constructing and interpreting data graphs and charts
- organizing analysis results
- making conclusions and recommendations based on data

Activity

Compiling the Data

After students collect their data, the next task is to compile the data in a way that allows each group access to all of the information. A natural tendency is to divide up the questionnaire into strips and give each small group all the answers to a single question, as was done with the Meals and Chores Survey in Investigation 3. The problem with this method is that it keeps students from seeing patterns that involve more than one variable, such as the relationship between age and where an injury took place. Rather, they need an organization similar to that of the Cat Cards, in which all the information about a single injury (comparable to all the information about a single cat) stays together, and students can sort injuries into piles to look at relationships in the data.

There are several ways to compile the class data. The starting point is a chart like the one shown on p. 99, with headings that reflect each of the questions on the survey. An option that leaves much of the work to students is to let them fill in their own data, on either a small or large form of the chart that you provide. A small chart can be passed around the room, with each student adding data in turn. Small charts have the advantage of being easily photocopied so that groups can cut apart the data into individual injuries to analyze it. If you use a large chart on the wall, students can come up one or two at a time and fill in their data.

If you prefer not to use class time to collate data, you can simply collect students' interview results at this time and transfer the data onto a chart yourself sometime before the next session. If you have access to a database program like Tabletop, enlist a couple of skilled student computer users to enter the data into the database. In any case, make a copy of the resulting data chart for each small group.

Whichever approach you take, begin by soliciting data from a few interviews—just enough for students to begin to work with, about 10–15 pieces of data. Record these on the board or on a large chart. While the rest of the data is being collated, students will be working in small groups with these preliminary data.

While we are getting all of our data in one place, you will start to work in your groups with this small amount of data. Try to answer a few questions with these data, and decide what you will want to do with the larger amount of data when it is ready. Experiment with a graph that you may want to use later with all the data. This is the time to plan how you can use the data to answer some of your questions.

One of the first issues likely to arise is that students will have to divide injuries into several categories. In one class, each small group came up with its own categories for injuries. One group classified them as *serious, medium,* or *not serious;* a second group used the categories *emergency room, doctor's office,* and *other.* A third group categorized the injuries as *broken bones, stitches,* or *other.*

During this period, students can begin to work on a categorization system. If data are being added to a large chart, they will be able to see if their system works as more data are added during the session. The **Dialogue Box,** Categories and Associations (p. 119), illustrates the process of building a set of categories for injuries. Other answers, such as where and when the injury occurred, will also need to be categorized, as in the list below.

Where they got hurt

cement ramp
field
tire swing
driveway
boardwalk
grassy field
rainbow bars
climbing structure

bars
climber
ramp
jungle gym
grass
front steps

Making Posters of Data Analyses

Before students begin to work with the data, refocus their thinking on the initial questions they listed in Session 1, as Issues to Study. Make sure that many of the questions get answered, perhaps by assigning a question to each group, or allowing each group to choose a question with the restriction that they cannot research a question that is already chosen.

Many of these questions probably involve relationships between variables. For example, the question "Do younger students get hurt on the swings more often than older ones?" requires making associations between the place the injury occurred and age of the student. Encourage students to go beyond simple bar charts of individual variables, since such graphs do not provide information about associations.

One way students might organize their inquiry is similar to the way they looked for associations when they used the Cat Cards. First they could cut apart their copy of the data chart into strips, one strip per injury. Then they could sort the data into piles by category. If they were working on the question about age as related to swing injuries, for example, they might make separate piles for younger and older students, then examine the frequency of swing injuries in each. A useful data representation in this case might be two bar charts of injury location, one for older students and one for younger students.

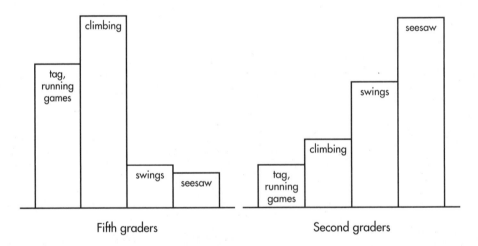

As students work, circulate and help them focus their data organization on the original class questions, making suggestions about data analysis techniques where you think it is appropriate. Remind students of the work they did with cats, looking for associations between, for example, gender and length. The **Teacher Note,** Taking Time for Data Analysis (p. 117), illustrates some of the representations that students may use in answering questions about the data.

If database software is available, this is the time for students to work with it to answer some of their questions.

Allow at least one and a half sessions for this data representation and analysis activity. Each group prepares a poster of their findings that has a combination of graphics and writing. The poster should include some numerical description of the data that involves fractions, decimals, or percents; for example, "60 percent of the injuries on the swings happened when no teacher was on the playground."

As students complete their posters, they display them around the room for other students to comment on. These posters will be part of the assessment for this unit. For the class discussion of the posters, each group writes one or two sentences about their findings on the board or chart paper. For example:

> Fifth graders get injured more on weekends than first or third graders.

As students work, you may have to remind them to concentrate on the mathematical content of the poster, rather than on the artistic decoration. Color can be an effective way to make graphs and charts communicate better, but too much decoration will distract both the students making the graph and the people who look at it for information.

Assessment

What We Recommend

After the posters are completed and all the students have had an opportunity to view them on the wall and to read each group's short summary of their findings, the class discusses what recommendations it would make based on all the findings. Show again the class list of Issues to Study from Session 1.

How would you answer these questions, based on what we found in our survey? Do you have something to say to each of these questions?

Spend some time exploring theories about the data. Challenge students to make statements based on the data. For example:

> I think we were right about age of the kids being important: 60 percent of the injuries to younger kids were on equipment, and only 35 percent of injuries to kids over 8 were on equipment.

Be sure to consider how the sampling process might have affected the data—if, for example, they didn't interview any students younger than second grade.

We've looked at the results now. What recommendations would you make that might address some of the problems you've identified? Let's make a list of those recommendations, and see if we agree about them.

Make a list with the class. Students' recommendations might relate to issues such as the following:

> The condition of playground equipment
>
> The placement of equipment, fields, and playing areas
>
> Adult supervision
>
> How crowded the playground should be
>
> Timing of recess; who goes to recess when
>
> How much playground activity should be formally organized

Base your assessment on a combination of the students' written work on the poster and their contributions to this discussion on safety recommendations. Consider the following:

> Does the poster ask and answer an appropriate and interesting question?
>
> Do the data answer the question?
>
> Are the data organized in a way that communicates to the audience?
>
> Are the data summarized in a few concise sentences?
>
> Are the charts and graphs accurate? appropriate? helpful to the reader?
>
> Has everyone in the group participated in making the poster?
>
> In the discussion, do students describe their data accurately?
>
> Are the students' conclusions based on the data?
>
> Are the students' statements and recommendations upheld by the data?

After students have finished making and discussing their recommendations, consider with them whom they could contact about their ideas. Is there any action that grows naturally out of these analyses? Encourage students to decide who needs to know about their findings, and who could do something about the recommendations. The **Teacher Note,** Making Final Presentations (p. 118), describes the options for reporting survey conclusions to appropriate groups; the extension Giving a Presentation (p. 116) has ideas for organizing such a report.

As the unit ends, you may want to use one of the following options for creating a record of students' work on this unit:

■ Students look back through their folders or notebooks and write about what they learned in the unit, what they remember most, and what was hard or easy for them. Students could do this work during their writing time.

■ Students select one or two pieces of their work as their best work, and you also choose one or two pieces of their work, to be saved in a portfolio for the year. You might include students' work on individual cats and their associations between variables for cats. Students' posters on playground injuries are an important component of their portfolios. Since students worked on these in groups, put the poster with one student's folder and put notes in the others' folders, telling where the poster is filed. Students can create a separate page with brief comments describing each piece of work.

■ You may want to send a selection of work home for families to see. Students write a cover letter, describing their work in this unit. This work should be returned if you are keeping a year-long portfolio of mathematics work for each student.

Sessions 3, 4, and 5 Follow-Up

If necessary, after Session 3 or 4, students work individually at home on parts of their group's presentation. In order to do this, the group will need to take a few minutes at the end of the period to divide up responsibilities. One possibility is that each student try to find a way to categorize data from the answers to a single question, such as where the injury occurred.

■ **More Research on Playgrounds** If students decide that they want to pursue playground safety further, encourage them to refine and develop their survey. They will need to decide whether they want more information from the sample they worked with or want to expand their sample to include more people or playgrounds. This data project could easily be expanded; the decision is yours and the students'.

■ **Giving a Presentation** See the **Teacher Note,** Making Final
Presentations (p. 118), for some specific suggestions about appropriate
audiences for an oral presentation. Students will need to decide ahead of
time what messages they want to convey and what visuals they want to
use to support their message. They may want to make more polished
graphs and charts, make overheads, or prepare a script and dramatize
some of their data. Usually, presentations about research have the following parts:

What questions were you studying?

Why are these questions important?

How did you study them?

What did you find?

Based on your findings, what recommendations do you have?

How do your findings compare to the national findings?

■ **Publishing Research Findings and Recommendations** The **Teacher
Note,** Making Final Presentations (p. 118), also contains suggestions for
places to publish research results. The questions listed above would then
be answered in a written report and submitted for publication.

Taking Time for Data Analysis

The playground investigation pulls together much of what the students have worked on during this unit. Students come up with questions to explore, design a survey, interview other students, collect and collate the data, and use the data to help answer their questions. Considering the complexity of the task, there is a relatively short time for students to complete their work. It is important that you keep the activity moving along through many points where it might get stuck.

Once students start discussing their personal experiences with playground injuries, they can go on for a long time. Similarly, students often have a hard time deciding on a set of questions, especially since they need to come to a consensus as a whole class. At these times, you will have to lead the class to some closure, even though students may not be quite ready to end the discussion.

It is most important that students have the opportunity to use their data to answer their original questions. This part of data investigations often gets short shrift because time is running out and students are coming to the end of their interest in the topic. Pace the investigation so that students have nearly two whole sessions to work with their data, to experiment with representations, and to discover what they have learned.

During the analysis phase, look for students who are ready to go beyond making graphs for individual variables and help them work on associations, as they did in the cats investigation. For example, if students have found that among the youngest children, 43 percent of the injuries are caused by falls, while this figure drops to 31 percent among third graders and 15 percent among fifth graders, they could show this finding with a bar graph or a chart.

A more complex example is comparing different types of injuries received by boys and girls. To show that, students could make a graph comparing the percent of boys and girls for each type of injury. The following is one such graph.

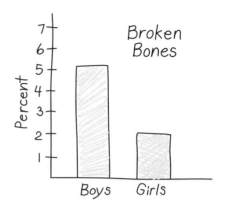

Another approach would be to calculate the total number of injuries separately for boys and girls, then figure out what percentage of injuries falls into each category. Below is such a graph, for girls, when the data are separated by gender.

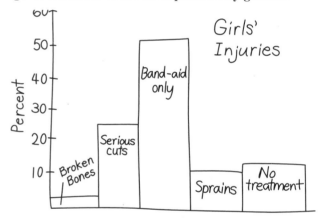

Many students want to make circle graphs. Two cautions are in order here: First, the calculations involved in making circle graphs may be difficult for students and distract them from the meaning of their data. (The "Percentractor" tool, introduced in the *Investigations* grade 5 unit *Name That Portion,* can simplify the process). Second, circle graphs of the data for two groups, such as boys and girls or older and younger students, can be more difficult to compare than the corresponding bar graphs. For comparing data, encourage your students to try using bar graphs and line plots.

When students have carefully researched a topic and end up with something to say, it is important that they find a good way of delivering their message. Following are some audiences to consider. Whatever audience you choose, desktop publishing programs can help your students generate formal, professional-looking reports with the computer. Students enjoy making things look "real," and teachers often find that the promise of real publication is a strong motivator.

Presentation to a Responsible Official If your students' work focuses on the school playground, their presentation might be directed to the principal and the superintendent of schools. If they have studied the broader community, you might do better with municipal officials. Usually, it's in the best interest of politicians to meet with schoolchildren, so they are likely to agree.

Newspaper The local newspaper may be interested in covering the story if your students have found something of general community interest. In one town, students found that the bike path was dangerous, especially for walkers. They studied the accidents reported by students along the bike path and sampled the area around the path to find out what kinds of accidents people reported. After they noticed that there were more problems for walkers than for riders, they suggested that the community construct a wide path alongside the bike path in order to keep the two kinds of traffic separate.

First they presented their ideas in the school, then to the school board, and finally to the newspaper. Members of the town government were interested and asked the class to make the presentation to them as well. With each of these presentations, students were careful to present a summary of their methods, their data, and their graphs, charts, and recommendations.

Local Television Your students may be interested in presenting a short segment on a local television station or on the school network. For such a presentation, they might practice in the classroom, videotaping a description of their project and making their graphic data displays large and easily readable. You might make a multimedia presentation if the technology is available to you.

School Library You may find that your students' work lends itself to the creation of a large book or booklet that can be donated to the school library for others to see. Making their work available to others on a long-term basis creates models of statistical research for younger students who may pursue such projects later. You may also find that the community library is interested.

Other Groups Students may be interested in describing their project and findings to the U.S. Public Interest Research Group, whose survey results they read about on Student Sheet 12, Danger on the Playground. Here's the address:

> U.S. PIRG Education Fund
> The National Association of State PIRGs
> 215 Pennsylvania Ave., S.E.
> Washington, DC 20003

Students might also search for civic groups and government offices that use the kind of information they have gathered to make policy.

Categories and Associations

While the class is collating all the data from their interviews on playground injuries (Compiling the Data, p. 110), small groups are starting to look at preliminary data and think about how they might organize it. The teacher looks on as Leon, Manuel, and Julie work together on this task.

Which question are you working on?

Leon: It's "Where did you get injured?" Here's our list of what kids said.

Manuel: It's hard, because people gave us different kinds of answers.

What do you mean?

Manuel: One kid said, "on the slide," another kid said, "under the trees," and another kid said, "next to the building."

Let's look at all of them and see if there are any that go together.

Julie: I think "under the trees" and "next to the building" go together, and they go with this one: "on the porch." They're all sort of general outside places.

Manuel: Yeah, none of those kids mentioned any equipment like the slide.

Julie: Maybe we could put together places that involve equipment and places that are general.

Leon: But what about the field? Lots of kids got injured on the field.

Manuel: Yeah, the field isn't a general place, it's a place where you play games.

Sounds like you have three categories now.

Julie: Right. Places where you play games, general places, and places with equipment.

That looks good. How are you going to show it?

Manuel: I think we should put each place someone said on a stick-on note. Then we can put the stick-on notes in the three categories.

Julie: Wait a minute! We thought that younger kids were going to get injured more on equipment and older kids in other places. How are we going to look at that?

Leon: We could write the kid's age on each stick-on note.

How are you going to see if the two things are related to each other?

Julie: Maybe we could make separate charts for kids under 8 and kids older than 8.

Leon: What will we do with 8-year-olds? One of the kids I interviewed was 8.

Julie: Let's do 8 and younger over here.

The Digits Game

Basic Activity

Students play a game in which they use randomly drawn digits to make numbers as close as possible to a target number.

As students make numbers from digits, they build their understanding of place value, relationships among the places, and the role of zero in a place value system. As they determine how far from the target the numbers they make are, they develop strategies for comparing numbers. Some students use mental addition or subtraction, others use written addition and subtraction, and others rely on numerical reasoning. Their work focuses on:

- reasoning about place value
- determining which of a set of numbers is closest to a target number
- developing strategies for comparing numbers

Materials

- Digit Cards (p. 153, 1 deck per playing group)
- Pencil and paper

Procedure

Step 1. Pick a target number.   Choose a number to use as a target number, for example, 1000.

Step 2. Randomly select the digits to be used. Mix the Digit Cards facedown and draw one for each place in the target number. For 1000, a group would draw four cards.

Step 3. Players arrange the digits.   Using paper and pencil, the players arrange any number of the digits shown on the cards drawn to make a number that is as close as possible to the target number. If the digits picked were 9, 4, 1, and 7, then 974 would be closest to the target number of 1000. 1479 would be the closest number *above* 1000.

Step 4. Share solutions.   Students compare different solutions to see which one comes closest to the target number.

Note: When this is done as a whole-class activity, it's more efficient for the teacher to draw the digits. Then students work individually to build a number. They can compare their solutions in a whole-class discussion or in small groups.

Variations

Large Target Numbers   To increase the level of difficulty, increase the size of the target number you are asking students to reach. For example, make a target number of 500,000, and have students draw 6 digit cards to make a number as close to the target as they can.

Largest or Smallest Number   With the digits they draw, students try to make the largest or smallest number they can (rather than a specific target number).

Largest or Smallest Sum   Students draw a given number of cards and create a two part addition problem with them, making the largest or the smallest sum possible. Decide on rules for the combinations of numbers. For example, with four digits drawn, students might be restricted to a two-digit plus two-digit addition problem, or perhaps a three-digit plus one-digit problem.

Suppose the digits drawn were 4, 3, 2, and 1. Then, if the rule was two-digit numbers, 41 + 32 (or 42 + 31) would give the largest sum possible—73. The problem 14 + 23 (or 13 + 24) would give the smallest sum—37. If the rule was a three-digit plus one-digit problem, then 432 + 1 would give the largest sum, and 123 + 4 the smallest.

Another variation would be to create two numbers that, when added, make a sum as close as possible to a target number (say, 100).

Largest or Smallest Difference   Following the procedures for Largest or Smallest Sum, students make subtraction problems.

Largest or Smallest Product   Following the procedures for Largest or Smallest Sum, students make multiplication problems.

Decimals   For these variations, add the decimal point card to the deck of Digit Cards.

Largest or Smallest Decimal   Students create the largest or smallest decimal number they can with a given set of digits. They must place the decimal point card to the left of their first digit.

Target Decimals   Students create a decimal number that is as close as possible to a target number. They may choose where they put the decimal point.

Sums, Differences, Products   Students create addition, subtraction, or multiplication problems as described above, but using decimal numbers. The same kinds of restrictions and variations can be applied.

One by One   For this variation, each student needs a set of Digit Cards. Set any goal (largest number, smallest number, close to a target number, and so forth) and tell students how many cards to draw. Students are to draw the cards one by one. As each card is turned over and before the next is flipped, the player must decide in which position to place that digit in order to have the best chance of reaching the goal.

Volume and Surface Area

Basic Activity

Students determine the number of cubes (volume) and the number of square stamps (surface area) in various cube configurations presented as diagrams on the overhead.

Students investigate ideas that are essential to learning about volume and surface area. The volume of a 3-D solid is generally measured in terms of the number of unit cubes that fit inside it. The surface of a solid is measured in terms of the number of unit squares it takes to cover it.

In order to count either cubes or squares, students must understand the structure of a 3-D object, and be able to count the cubes that fit inside the object or the squares that cover it. Students also learn that 3-D objects with the same volume do not necessarily have the same surface area. Students focus on:

- understanding the structure of a 3-D object (for instance, how its faces fit together)
- organizing and finding the number of cubes that fill simple 3-D solids, such as rectangular prisms
- organizing and finding the number of squares that cover simple 3-D solids, such as rectangular prisms
- recognizing that 3-D objects with the same volume do not necessarily have the same surface area

Materials

- Overhead projector
- Overhead transparencies of the Cube Diagrams (p. 154). Copy the transparency master and cut apart the cube diagrams. Include the numbers to help orient the diagram properly on the overhead. Store each lettered set of five diagrams in its own envelope, as these build on one another and should be done in sequence, if possible.
- Interlocking cubes (40–50 per pair)
- Paper and pencil

Procedure

Step 1. Introduce the "square stamp" idea. When doing this activity for the first time, show students a cube and ask them how many square stamps it takes to completely cover a single cube if the stamp is the same size as the square faces of the cube. Let students examine a cube if they need to. You may want to sketch the following on the overhead or board. If students have done the activity before, skip Step 1.

Square stamp   Cube

Step 2. Display a cube diagram on the overhead. Students work with a partner to figure out how many stamps it would take to cover it, and how many cubes it would take to build it. Have cubes available for students to figure or check with. After they have had some experience with this task, encourage them to figure out answers by looking only at the diagrams. (Of course, they can still check by using cubes.)

Step 3. Students share their strategies. Encourage students to look for patterns in the sequence of cube configurations you present and to explain numerical patterns in terms of spatial organizations.

For instance, the number of stamps needed for B-3 is just 6 more than the number needed for B-2 because it has a single band of squares added (2 on the top, 2 on the bottom, 1 on each side), as shown below.

 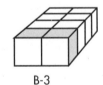

B-2   B-3

Note that adding two cubes to B-2 does not add two faces (stamps) on the end, because the original endfaces of B-2 are covered by the new cubes.

Variations

Same Volume, Different Surface Area

■ Begin with any 3-D diagram. Students find the volume (number of cubes) and the surface area (number of stamps) for that figure, then rearrange the cubes to make a rectangular solid with the same volume but different dimensions and a different surface area.

■ Students begin with 20 cubes. They make a shape (using all 20 cubes) that has a small surface area; then make another shape with the 20 cubes that has a larger surface area. What is the largest surface area they could make by putting together the 20 cubes?

Outline and Dimensions Only   On the overhead, sketch the outline of a diagram for a rectangular solid, without the interior lines that show where the cubes connect. Label the dimensions. For example:

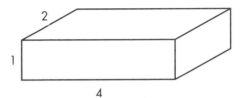

Students work together to find the number of cubes in this figure and the number of square stamps needed to cover it entirely.

Dimensions Only   For an extra challenge, give only the dimensions—no visual image. For example, for the solid above: 1 by 2 by 4. How many cubes are needed? How many square stamps are needed to cover its surface?

The following activities will help ensure that this unit is comprehensible to students who are acquiring English as a second language. The suggested approach is based on *The Natural Approach: Language Acquisition in the Classroom* by Stephen D. Krashen and Tracy D. Terrell (Alemany Press, 1983). The intent is for second-language learners to acquire new vocabulary in an active, meaningful context.

Note that *acquiring* a word is different from *learning* a word. Depending on their level of proficiency, students may be able to comprehend a word upon hearing it during an investigation, without being able to say it. Other students may be able to use the word orally, but not read or write it. The goal is to help students naturally acquire targeted vocabulary at their present level of proficiency.

We suggest using these activities just before the related investigations. The activities can also be led by English-proficient students.

Investigation 1

seconds, minutes, score, compare, range

1. Point to the second hand on a clock. Count the passage of 20 seconds. Then challenge students to draw as many circles as they can in 20 seconds.

2. After 20 seconds, students count the number of circles they drew. Explain that their score will be the number of circles drawn.

3. Record students' scores next to their names on the board. Then ask questions that focus on comparing the scores.

   Let's compare these scores. Is Duc's score higher or lower than Sofia's score? Who has the lowest score? Who has the highest score?

4. Write the lowest and highest score on the board. Ask students to count how many scores fall between these two scores. Start a line plot with values from the lowest to the highest scores. Explain that this is the range for these scores.

5. Repeat the activity, but this time challenge students to draw circles for 1 minute. Ask them to predict whether their scores will be 3 times their first scores (since they now have 3 times the amount of time to draw), or if other factors will change the outcome.

Investigation 5

emergency room, injury, serious

1. Show a picture of a hospital emergency room. Explain that people go to this part of the hospital when they need a doctor right away for a serious injury.

2. Have volunteers enact the following injuries: (1) unconsciousness resulting from a bike accident; (2) a splinter in a finger; (3) a broken arm from a fall from climbing equipment; (4) a broken leg from a fall from a skateboard; (5) a scrape on the arm from running and falling; (6) a bruised arm from being hit by a fast-moving ball. After each enactment, identify the injury, and ask the group if the person needs to go to the emergency room.

3. Explain that we might group these injuries under two headings: *Serious* and *Not serious*. After you write these headings on the board, give an example of a serious injury that the students enacted. Record this (with a simple sketch) under the Serious heading. Challenge students to group the rest of the injuries under these two headings.

Blackline Masters

_____, 19____

Dear Family,

In mathematics, our class is starting a unit on data. The data we study will address questions ranging from "How long can you stand on one foot?" to "How safe is our playground?"

Much of the data we read and hear about every day involves comparisons—of everything from automobiles to cold remedies. A lot of the data we see on TV and in newspapers have something else in common: They are based on samples. For example, a survey of a small number of people may be taken to represent the nation as a whole. Your child will be learning about both processes—making comparisons of data, and taking samples. There are many ways you can help.

First, your child will be collecting "foot balancing" data from you and other adults. Children have fun with this topic while they learn about collecting data. They will compare the adult data with data on themselves, and try to decide who balances better, the adults or themselves.

Later, the children will work on ways to select a reasonable sample. One way they do this is by sampling the pages of daily newspapers to figure out the fraction of the paper that is advertising. Share pages from a newspaper at home and listen to your child's approach to this process.

Our final investigation looks at playground safety. Each year, more than 170,000 playground injuries require emergency room treatment. Our class will conduct a safety study, designing a survey and interviewing students of different ages. Using the data we gather, the class will make some recommendations about our playground. Be sure to discuss this study with your children—it is an important piece of research for the school community.

As you read or hear statistics, remark on them to your child. For example, if you hear that 1 out of 4 doctors prefers a product, you might say, "I wonder how many doctors they asked?" or, "If more doctors were sampled, would they get the same findings?" Questions like these encourage children to think about data, to question data, and to make better decisions based on the data they hear or read every day.

Sincerely,

Collecting Data on Balancing

For the balancing test, follow this procedure.

■ Person is allowed to get comfortably balanced on one foot before closing eyes.

■ Timing starts when person closes eyes and says "go."

■ Person can wiggle in place, but not hop or spin. Some part of the foot on the floor must always touch the floor.

■ Foot that is up can't touch the floor or a wall or piece of furniture.

■ Four things can end the test:

Person puts foot down.

Person opens eyes.

Person hops or touches an object for balance.

Person balances for 3 minutes.

■ Person gets one practice trial for each foot.

■ If the person is still balancing at 3 minutes, stop the test and record 3 minutes as the time.

When you are collecting data, be sure to have your eyes on the clock or watch before the person says "go." Right after each test, record two things:

The length of time

Which foot the person balanced on

How Long Do Adults Balance?

How long can adults balance on each foot with their eyes closed? Test two or more adults. Record your findings. Follow the same rules you used at school. The person can practice first, one practice only, on each foot. Remember that 3 minutes is the longest time you can record.

Name of adult	Balance time on right foot	Balance time on left foot

Do you think the adults will be different from you and your classmates? How might they be different? Write two predictions.

1.

2.

Student and Adult Balancers

0

Student Balancing Time on Right Foot

0

Student Balancing Time on Left Foot

0

Adult Balancing Time on Right Foot

0

Adult Balancing Time on Left Foot

Who Are Better Balancers?

These statements compare students' balancing data with adults' data. Read each statement. Decide who are better balancers, according to that statement, and write *adults* or *students* in the blank. Write a reason for your answer.

1. Almost $\frac{3}{4}$ of the students balanced for 30 seconds or more, but only $\frac{1}{2}$ of the adults did.   _____

2. While $\frac{2}{3}$ of the adults are above 45 seconds, only $\frac{1}{2}$ of the students are above 45 seconds.   _____

3. The median for students is 40 seconds, and it's only 30 seconds for adults.   _____

4. The students had the lowest score (2 seconds), and they had a larger fraction of scores under 10 seconds. Only $\frac{1}{20}$ of the adults did less than 10 seconds, but $\frac{1}{8}$ of the students did less than 10 seconds.   _____

5. Two students balanced more than 1 minute, and no adults did. Also, the top $\frac{1}{4}$ of students scored 50 seconds or more, and the top $\frac{1}{4}$ of adults scored only 40 seconds or more.   _____

6. Choose one of the statements above. Make two line plots—one for adults and one for students—that fit the statement. You can use the back of this sheet.   _____

Mystery Balancers Data

Mystery Balancers A

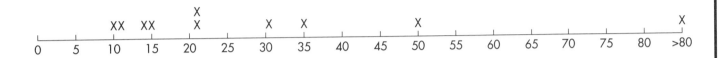

Mystery Balancers B

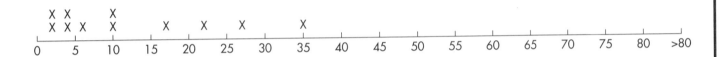

Mystery Balancers C

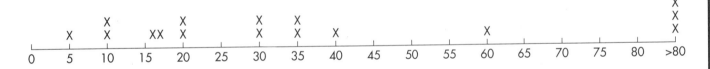

Mystery Balancers D

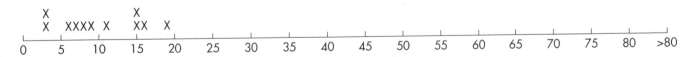

Writing About the Mystery Balancers

Circle the letter of your mystery data: A   B   C   D

Write a report that answers these questions:

- How are your mystery balancers similar to the students in your class? How are they different?

- Is your class better or worse at balancing than your mystery balancers? Or is there no clear "better" group? Give reasons for your answer.

- Which group of mystery balancers do you think you have? Why? If you think two groups are equally likely, name the groups and tell why.

Remember to make statements based on the data.

The Mystery Groups

Gymnasts, ages 9–20

Karate students, ages 16–49

First and second graders, ages 6–8

People over 50

Collecting Cat Data

Cat's Name		
Gender	Age (years)	Weight (pounds)
Body length (inches)		Tail length (inches)
Fur color	Eye color	Pad color
Other		

Cat's Name		
Gender	Age (years)	Weight (pounds)
Body length (inches)		Tail length (inches)
Fur color	Eye color	Pad color
Other		

Lady Jane Grey

Gender: female
Age: 4 years
Weight: 8.5 pounds
Body length: 19 inches
Tail length: 11 inches
Fur color: gray
Eye color: yellow
Pad color: gray

Peau de Soie

Gender: female
Age: 15 years
Weight: 7 pounds
Body length: 16 inches
Tail length: 13 inches
Fur color: orange, black, and white
Eye color: green
Pad color: pink

Other: Peau de Soie means "skin of silk" in French; her nickname is Peau (rhymes with *go*).

Data: Kids, Cats, and Ads

Mittens

Gender: female
Age: 14 years
Weight: 10.5 pounds
Body length: 17 inches
Tail length: 11 inches
Fur color: orange and white
Eye color: yellow
Pad color: pink

Other: Mittens has six toes on each foot.

Tigger

Gender: female
Age: 4 years
Weight: 8 pounds
Body length: 17 inches
Tail length: 10 inches
Fur color: orange, black, and white
Eye color: yellow
Pad color: brown

Weary

Gender:	male
Age:	8 years
Weight:	15 pounds
Body length:	17 inches
Tail length:	12 inches
Fur color:	black and white
Eye color:	green
Pad color:	pink

Ravena

Gender:	female
Age:	6 years
Weight:	14 pounds
Body length:	23 inches
Tail length:	12 inches
Fur color:	orange, black, gold, and white
Eye color:	yellow
Pad color:	pink and black

Data: Kids, Cats, and Ads

Lady

Gender:	female
Age:	10 years
Weight:	8.5 pounds
Body length:	17 inches
Tail length:	13 inches
Fur color:	gray, brown, and white stripes
Eye color:	yellow
Pad color:	black

Wally

Gender:	male
Age:	5 years
Weight:	10 pounds
Body length:	18 inches
Tail length:	12 inches
Fur color:	black and white
Eye color:	green
Pad color:	pink and black

Other: Wally is the brother of Peebles.

Oddfuzz

Gender:	male
Age:	5 years
Weight:	18 pounds
Body length:	21 inches
Tail length:	9 inches
Fur color:	orange and white
Eye color:	yellow
Pad color:	pink

Melissa

Gender:	female
Age:	8 years
Weight:	11 pounds
Body length:	21 inches
Tail length:	11 inches
Fur color:	white, black, and orange
Eye color:	yellow
Pad color:	pink

Data: Kids, Cats, and Ads

Peebles

Gender: female
Age: 5 years
Weight: 9 pounds
Body length: 17 inches
Tail length: 11 inches
Fur color: gray
Eye color: green
Pad color: black

Other: Peebles is Wally's sister.

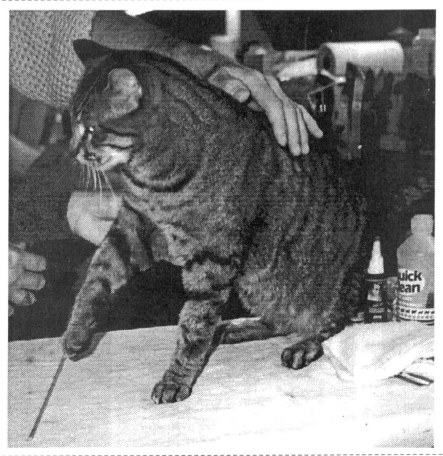

K.C.

Gender: male
Age: 5 years
Weight: 16 pounds
Body length: 24 inches
Tail length: 12 inches
Fur color: brown and black stripes, some white
Eye color: yellow
Pad color: black

Pepper

Gender: male
Age: 2 years
Weight: 12 pounds
Body length: 17 inches
Tail length: 9 inches
Fur color: orange
Eye color: yellow
Pad color: pink

Other: Pepper was known as an escape artist at the animal shelter where he was living.

Strawberry

Gender: female
Age: 16 years
Weight: 14.5 pounds
Body length: 21 inches
Tail length: 10 inches
Fur color: gray, brown, and white stripes
Eye color: green
Pad color: black

Data: Kids, Cats, and Ads

Alexander

Gender: male
Age: 18 years
Weight: 11 pounds
Body length: 21 inches
Tail length: 11 inches
Fur color: brown and
                 black stripes,
                 some white

Eye color: green
Pad color: black

Other: Alex's favorite foods are vanilla ice cream and bacon, which he will steal off the table.

Misty

Gender: male
Age: 1 year
Weight: 9 pounds
Body length: 18 inches
Tail length: 11 inches
Fur color: gray, white,
                 and black
Eye color: green
Pad color: pink and
                 black

Data: Kids, Cats, and Ads

George

Gender:	male
Age:	12 years
Weight:	14.5 pounds
Body length:	21 inches
Tail length:	13 inches
Fur color:	black and white
Eye color:	green
Pad color:	black

Diva

Gender:	female
Age:	3.5 years
Weight:	11 pounds
Body length:	20 inches
Tail length:	12 inches
Fur color:	gray, black, brown stripes with white patches
Eye color:	green
Pad color:	pink

Data: Kids, Cats, and Ads

Gray Kitty

Gender: female
Age: 3 years
Weight: 9 pounds
Body length: 15 inches
Tail length: 8.5 inches
Fur color: gray
Eye color: green
Pad color: gray

Other: Gray Kitty was living at an animal shelter.

Tomodachi Joto

Gender: male
Age: 1 year
Weight: 6.5 pounds
Body length: 14 inches
Tail length: 1.5 inches
Fur color: white and red
Eye color: gold
Pad color: pink

Other: Tomodachi Joto means "best friend" in Japanese. Nicknamed Joto, he is a Japanese bobtail cat.

Harmony

Gender:	male
Age:	3 years
Weight:	12 pounds
Body length:	24 inches
Tail length:	11 inches
Fur color:	black
Eye color:	greenish gold
Pad color:	black

Augustus

Gender:	male
Age:	2 years
Weight:	10 pounds
Body length:	21 inches
Tail length:	11 inches
Fur color:	black and white
Eye color:	yellow, green, blue
Pad color:	pink and black

Other: Augustus, Gus for short, is a long-haired cat. He was found as a stray.

Data: Kids, Cats, and Ads

Charcoal

Gender:	male
Age:	11 years
Weight:	12 pounds
Body length:	21 inches
Tail length:	13 inches
Fur color:	black and white
Eye color:	yellow
Pad color:	black

Other: Charcoal has big feet.

Cleopatra

Gender:	female
Age:	4 years
Weight:	7 pounds
Body length:	18 inches
Tail length:	9 inches
Fur color:	black and white
Eye color:	golden green
Pad color:	pink

Finding Familiar Fractions

During class, write down the questions. Fill in the numbers of people who answered *yes* and *no,* and the total number in your class today. Do the rest for homework. Use either Data Strips or numerical reasoning to find familiar fractions.

Question _____

Answer	Number who answered	Total in class	Fraction who answered	Familiar fraction
Yes				
No				

Question _____

Answer	Number who answered	Total in class	Fraction who answered	Familiar fraction
Yes				
No				

Question _____

Answer	Number who answered	Total in class	Fraction who answered	Familiar fraction
Yes				
No				

Data: Kids, Cats, and Ads

Small-Group Sampling

1. Question _____

_____

2.

Results of small-group sample	
Response	Number of students

3.

Prediction for the whole class, based on sample		
Response	Number of students	Fraction of class

Give a reason for your prediction.

4.

Prediction for the whole class, based on sample		
Response	Number of students	Fraction of class

5. Compare your sample results to the population results. Tell how your small group was similar to or different from the whole class.

Meals and Chores Survey

---

1. When your family eats dinner, do you all eat together at the same time?

    yes   sometimes   no

---

2. Do you watch television while you are eating dinner?

    yes   sometimes   no

---

3. What's your favorite meal of the day—breakfast, lunch, or dinner?

    breakfast   lunch   dinner

---

4. Do you get to choose what you eat for dinner?

    yes   sometimes   no

---

5. Do you help set the table?

    yes   sometimes   no

---

6. Do you help cook?

    yes   sometimes   no

---

7. If you had a choice of cooking dinner, setting the table, or cleaning the dishes, which would you choose?

    cook dinner   set table   clean dishes

Questions are from a national survey,"America's Children Talk About Family Time, Values, and Chores," sponsored in 1994 by the Massachusetts Mutual Life Insurance Co.

Survey Results

Responses	National sample	Our class

1. When your family eats dinner, do you all eat together at the same time?

Responses	National sample	Our class
yes	82%	
sometimes	8%	
no	9%	

2. Do you watch television while you are eating dinner?

Responses	National sample	Our class
yes	29%	
sometimes	19%	
no	53%	

3. What's your favorite meal of the day—breakfast, lunch, or dinner?

Responses	National sample	Our class
breakfast	24%	
lunch	33%	
dinner	43%	

4. Do you get to choose what you eat for dinner?

Responses	National sample	Our class
yes	27%	
sometimes	34%	
no	40%	

5. Do you help set the table?

Responses	National sample	Our class
yes	57%	
sometimes	19%	
no	24%	

6. Do you help cook?

Responses	National sample	Our class
yes	26%	
sometimes	37%	
no	37%	

7. If you had a choice of cooking dinner, setting the table, or cleaning the dishes, which would you choose?

Responses	National sample	Our class
cook dinner	37%	
set table	44%	
clean dishes	18%	

Data (rounded to nearest percent) from "America's Children Talk About Family Time, Values, and Chores," 1994, Massachusetts Mutual Life Insurance Co.

Data: Kids, Cats, and Ads

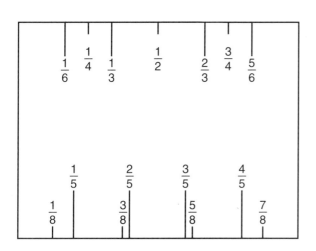

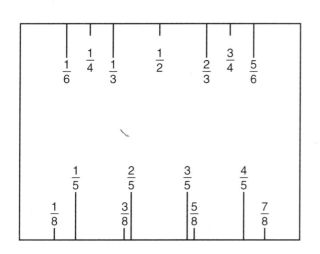

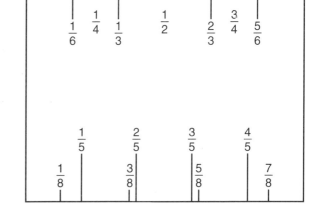

Investigation 4 Resource

Data: Kids, Cats, and Ads ▪ **151**

Danger on the Playground

A survey of playground injuries in the United States was done in 1994. Each year 170,000 children are seriously injured on playgrounds. The survey involved 443 playgrounds in 22 states. The researchers defined an injury as anything serious enough that the person had to go to the emergency room. They found these problems:

- In 92% of the playgrounds, the surfaces under equipment were not soft enough. Three-fourths of all injuries are caused by falls. Many of these falls cause head injuries, which can be very serious.

- In 57% of the playgrounds, the climbers and slides were too high—over 10 feet tall. Slides account for about 30% of all playground injuries. Mostly, children fall from the top or sides of the slide.

- Climbers account for 41% of injuries to school children. Many of these injuries involve broken bones.

- In 76% of playgrounds, the swings were too close to each other and to other equipment. About 28% of injuries are caused by swings.

- A serious cause of injuries to young children is getting their heads stuck in equipment. The survey found that 55% of the playgrounds had equipment on which children could get their heads stuck between the rungs.

- About 40% of all playground injuries involve children 6 to 8 years old. Another 40% involve preschool children. About 20% involve children older than 8.

Reference: "Playing It Safe: A Second Nationwide Safety Survey of Public Playgrounds," U.S. Public Interest Research Group and Consumer Federation of America, May 1994.

1	2	3	4
5	6	7	8
9	0	0	0

•

Decimal card
for decimal variations

A–1	B–1	C–1	D–1

A–2	B–2	C–2	D–2

A–3	B–3	C–3	D–3

A–4	B–4	C–4	D–4

A–5	B–5	C–5	D–5

Data: Kids, Cats, and Ads